International Mathematics

Coursebook 3

International Mathematics

Coursebook 3

Andrew Sherratt

HODDER
EDUCATION
AN HACHETTE UK COMPANY

Hachette UK's policy is to use papers that are natural, renewable and recyclable products and mode from wood grown in well-managed forests and other controlled sources. The logging and manufacturing processes are expected ta conform to the environmental regulations of the country of origin.

Orders: please contact Hachette UK Distribution, Hely Hutchinson Centre, Milton Road, Didcot, Oxfordshire, OX11 7HH. Telephone: +44 (0)1235 827827. Email education@hachette.co.uk. Lines are open from 9 a.m. to 5 p.m., Monday to Friday. You can also order through our website: www.hoddereducation.com

© Andrew Sherratt 2009
First published in 2009 by
Hodder Education, an Hachette UK Company,
Carmelite House, 50 Victoria Embankment
London EC4Y 0DZ

Impression number   10   9   8
Year   2022

Cover photo © PhotoAlto Agency RF/Punchstock
Illustrations by Macmillan Publishing Solutions
Typeset in 12.5 pt/15.5 pt Garamond by Macmillan Publishing Solutions

Printed and bound by CPI Group (UK) Ltd, Croydon, CR0 4YY

A catalogue record for this title is available from the British Library

ISBN 978 0 340 96744 7

Contents

Graphs of linear equations

In Unit 2 of Coursebook 2 we learned how to solve equations that have one variable, such as:

$$4x + 8 = 20 \qquad \text{to give} \qquad x = 3$$

Equations like this with one variable have only one solution. That means there is only one value for x that will make the equation true.

Now we will look at equations that connect two different variables, x and y, such as:

$$y = 3x + 2$$

This equation connects each different value of x with only one value of y, and also in this case each different value of y is connected with only one value of x.

For example: If $x = 1$, then $y = (3 \times 1) + 2 = 5$
and if $y = 1$, then $1 = 3x + 2$, $3x = -1$ and $x = -\frac{1}{3}$

For any value of x we can think of, we can find the value of y that is connected with it, and vice versa, but all these pairs of x- and y-values will be true only for the equation $y = 3x + 2$.

This kind of equation connecting two variables is also called a **function**.
We say that y is a function of x and we can write this as fx.

We have already learned about x- and y-values connected together in pairs like this. We called them coordinates and we showed that they tell us exactly where one point is on the Cartesian plane. (Review the work in Unit 11 of Coursebook 1 on the Cartesian plane and plotting coordinates.)

For the equation $y = 3x + 2$, if we choose some values for x and work out the value of y that is connected with each of them, we obtain pairs of values. We can plot these pairs of numbers (coordinates) on a Cartesian plane and draw the graph for this equation by joining the points with a line.

A Drawing a graph from an equation

If we know the equation that connects the y-values with the x-values, we can use this to draw a graph for this equation.

Sometimes, we are told which values of x to use to plot the graph, and we can use the equation to work out the corresponding y-values for each of these x-values.

If values for x are not given in a question, you must choose your own.

Choosing x-values for points to plot

- Always choose **at least three** (3) values for x, as we need to plot a minimum of three points for a graph to be accurate. It is a good idea to choose five values of x if possible.
- The value of $x = 0$ should always be one of the values you choose, if possible. We will learn later that the point where $x = 0$ is an important point. It is where the graph crosses the y-axis.
- Do not choose all the values for x too close together. Your graph will be more accurate if the points plotted are reasonably spaced from each other, say at least 1 or 2 units.

Example

Draw (or plot) the graph of the equation $y = 2x + 3$.

We are not told what x-values to use in this question, so we will choose our own.
Let's choose $x = -2$, $x = -1$, $x = 0$, $x = 1$ and $x = 2$.
Now we must now work out the values for y for each of these.

If $x = -2$
$$y = (2 \times -2) + 3$$
$$= -4 + 3$$
$$= -1$$

If $x = 0$
$$y = (2 \times 0) + 3$$
$$= 0 + 3$$
$$= 3$$

If $x = 2$
$$y = (2 \times 2) + 3$$
$$= 4 + 3$$
$$= 7$$

If $x = -1$
$$y = (2 \times -1) + 3$$
$$= -2 + 3$$
$$= 1$$

If $x = 1$
$$y = (2 \times 1) + 3$$
$$= 2 + 3$$
$$= 5$$

Write these values in a table to show how they are connected:

x	−2	−1	0	1	2
y = 2x + 3	−1	1	3	5	7

Or write the values as the coordinates of the points we will plot on the graph:

$(-2, -1)$, $(-1, 1)$, $(0, 3)$, $(1, 5)$ and $(2, 7)$

● Label each of the points that you plot with its coordinates.
● Label the graph with the equation that it shows.

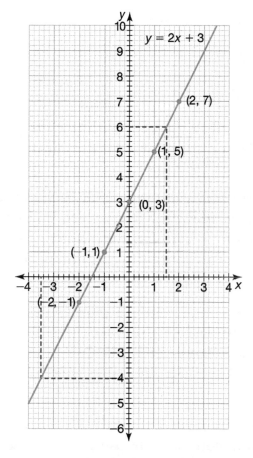

Once we have drawn the graph for an equation, we can use it to find out the value of y for any value of x, and vice versa.

● Start on the x-axis at the value of x you are looking at.
● Use a ruler to draw a line straight up (or down) until you meet the line of the graph.
● Now use the ruler to draw a line to the left (or right) until it meets the y-axis.
● Read the value of y from the scale on the y-axis.

If you start with a y-value, then you need to follow the steps in reverse.

We can also find these values by substituting the value of one of the variables into the equation and solving it to find the value of the other variable. However, using the graph can be much easier as we are just 'reading' the value from the graph.

Example

From the graph of the equation $y = 2x + 3$ on page 3, find these values.

i) The value of y when $x = 1.5$

When $x = 1.5$, $y = 6$.

ii) The value of x when $y = -4$

When $y = -4$, $x = -3.5$.

Choosing a scale for the graph

Graph paper is divided into large squares, and each of these large squares is also divided into a number of smaller squares. This grid of squares helps us to plot the points accurately.

However, we must choose the value that each of these squares (big and small) will show on the x-axis and also on the y-axis. This is called the **scale** of the graph. The scale we choose will determine if the graph is large (uses a lot of the graph paper) or small (uses only a small part of the graph paper).

The graph drawn in the example on page 3 uses this scale:
x-axis: 1 large square stands for 1 unit
y-axis: 1 large square stands for 1 unit.
So, in this case, we have used the same scale for the x-axis and the y-axis.

Sometimes you will be given a scale to use when you plot a graph, but most of the time you must choose your own scale. Here are some tips to help you.

1 Look at the coordinates of all the points you want to plot. Find the smallest and the biggest value of the x-coordinates. This means that the x-axis must be able to **show all numbers from the smallest x-coordinate to the biggest one**.
2 Choose a scale for the x-axis (such as 1 large square stands for 1 unit, or 2 units, or 4 units, or 5 units, or 10 units, etc.) so that all the points you are going to plot will be spread across most of your graph paper.
3 Repeat steps 1 and 2 for the y-coordinates of the points you want to plot, and choose a scale for the y-axis.

4 The scale used for the x-axis does not need to be the same as the scale used for the y-axis. Often we do use the same scale for both the x- and y-axes, but it will not always be the best choice.

5 Choose a scale for each of the axes so that the graph you want to draw will cover more than half of the graph paper you are using. The bigger the graph you draw, the more accurate the results you get from the graph will be.

 Exercise 1

1 For each of the eight equations below:

i) choose five values of x to plot and work out the value of y for each of these x-values

ii) write down the coordinates of the five points you will plot

iii) choose a scale for the x-axis and one for the y-axis

iv) plot your five points and join them with a straight line.

Plot these four graphs on the **same** Cartesian plane.

a) $y = x$ b) $y = x + 1$ c) $y = x - 1$ d) $y = x + 2$

What is special about these four graphs?

Plot these four graphs on the **same** Cartesian plane.

e) $y = -x$ f) $y = -x + 1$ g) $y = -x - 1$ h) $y = -x + 2$

What is special about these four graphs?

2 Choose a suitable scale and plot five points to draw the graph of the equation $y = 4x - 1$.
Use the graph to find:

a) the value of y when $x = -0.5$, $x = 1.6$, $x = 2.5$
b) the value of x when $y = -3.4$, $y = 7.8$, $y = 12.2$.

3 Choose a suitable scale and plot five points to draw the graph of the equation $y = 4 - x$.
Use the graph to find:

a) the value of y when $x = -2.5$, $x = 0.4$, $x = 5.2$
b) the value of x when $y = -4.5$, $y = 4.5$, $y = 8.8$.

B Gradient (also known as slope)

The **gradient** or **slope** of a hill measures how 'steep' the hill is. It is defined as the ratio of the distance up to the distance along:

$$\text{gradient} = \frac{\text{distance up}}{\text{distance along}}$$

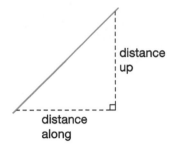

The same is also true for any straight line. To find the gradient of a straight-line graph, we need to draw a right-angled triangle to measure the distance up (**vertical distance** between the points), and the distance along (**horizontal distance** between the points).

For any straight line:

$$\text{gradient} = \frac{\text{vertical distance}}{\text{horizontal distance}}$$

The gradient of a line can be positive, zero, negative or *undefined*:

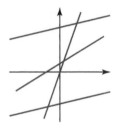

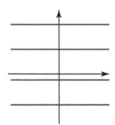

Positive gradients go 'uphill' from left to right.

Zero gradients are 'flat' from left to right.

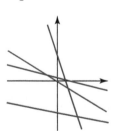

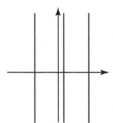

Negative gradients go 'downhill' from left to right.

Undefined gradients go 'straight up'.

Here is a general rule for working out the gradient of **any** line. Choose any two points $A(x_2, y_2)$ and $B(x_1, y_1)$ on the line:

The vertical length between the points will be:

$$AH = AL - HL$$
$$ = AL - BK$$

But remember that the y-coordinate of a point tells us the distance of the point from the x-axis. So y_2 is the same as AL and y_1 is the same as BK and

$$AH = y_2 - y_1$$

In the same way, the horizontal length between the points will be:

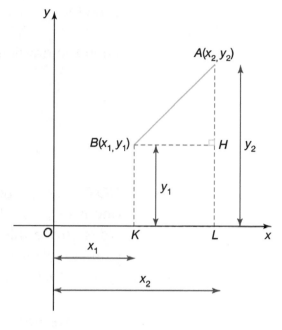

$$BH = OL - OK$$

and the x-coordinate of a point tells us the distance of the point from the y-axis. So x_2 is the same as OL and x_1 is the same as OK and

$$BH = x_2 - x_1$$

So the gradient of the line segment AB is given by

$$\text{gradient} = \frac{\text{vertical distance}}{\text{horizontal distance}}$$
$$= \frac{AH}{BH} = \frac{y_2 - y_1}{x_2 - x_1}$$

From this we can see that a different way to describe the gradient of a line is:

$$\text{gradient} = \frac{\text{change in } y\text{-coordinate}}{\text{change in } x\text{-coordinate}}$$

The **gradient** of a line joining two points (x_1, y_1) and (x_2, y_2) is equal to $\dfrac{y_2 - y_1}{x_2 - x_1}$.

Example

Find the gradient of the line joining the points $P(-2, 1)$ and $Q(5, 8)$.

First, we must decide which point is which, so let P be (x_1, y_1) and Q be (x_2, y_2).

Then the gradient of $PQ = \dfrac{y_2 - y_1}{x_2 - x_1}$

$$= \dfrac{8 - 1}{5 - (-2)}$$

$$= \dfrac{7}{7} = 1$$

NOTE: It does not matter which point you say is (x_1, y_1) and which one is (x_2, y_2). The answer for the gradient will be the **same** both ways. Prove this for yourself by calculating the gradient in the example above again, but this time let $P(-2, 1)$ be (x_2, y_2) and $Q(5, 8)$ be (x_1, y_1).

We need only **two** of the points on a line to be able to calculate **the gradient for the whole line.**

Exercise 2

1 Calculate the gradient of the line segment joining each of these pairs of points.

a) $A(5, 1)$ and $B(1, 4)$ b) $C(3, -1)$ and $D(-1, 2)$
c) $E(3, 9)$ and $O(0, 0)$ d) $F(5, -2)$ and $G(3, 1)$
e) $H(4, 6)$ and $J(-3, 2)$ f) $K(2, 4)$ and $L(-2, -2)$

2 Calculate the gradient of each of these lines.

a)
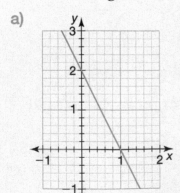
b)
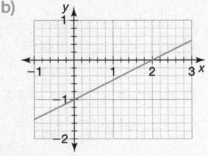
c)
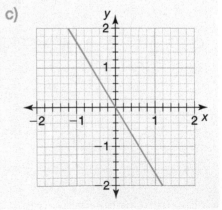

3 Here are the coordinates of three points on a straight line: $X(7, 5)$, $Y(-1, 1)$ and $Z(-3, 0)$.
Calculate the gradient of:

a) the line segment XY
b) the line segment YZ
c) the line segment XZ.
d) Use gradient calculations to show that the point $A\left(2, 2\frac{1}{2}\right)$ is also on the straight line XYZ.

4 Here are the graphs of some straight lines.
Decide if the gradient of each line is positive or negative. Do **not** calculate the gradient.

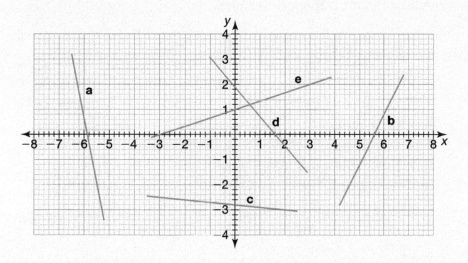

C Horizontal and vertical lines

Here is a graph of a **horizontal** line.

If we choose two points on this line, such as $P(2, 3)$ and $Q(-1, 3)$, we can calculate the gradient of the line:

$$\text{gradient of } PQ = \frac{y_Q - y_P}{x_Q - x_P}$$

$$= \frac{3 - 3}{-1 - 2}$$

$$= \frac{0}{-3} = 0$$

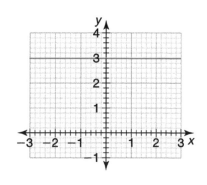

We find that the gradient of this horizontal line is 0. This makes sense as a horizontal line does not go 'up' or 'down' at all – it is completely 'flat'.

You will notice that the y-coordinate of **every** point on this horizontal line is the same (3). So we can say that the equation of this line is $y = 3$. For every value of x, the value of y is always 3.

We can draw many other horizontal lines on our Cartesian plane. Each horizontal line will contain points that all have the same y-coordinate.

In general, the equation for any horizontal line will be $y = c$, where c is the constant number on the y-axis that the line passes through (the y-coordinate of each point on the line).

Calculate the gradient for some other horizontal lines and you will find that it is always 0.

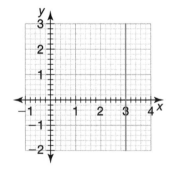

- The gradient of a horizontal line is **0**.
- The equation for a horizontal line is $y = c$, where c is the constant number on the y-axis that the line passes through.
- Horizontal lines are always **parallel to the x-axis**.

Here is a graph of a **vertical** line.

If we choose two points on this line, such as $E(3, 2)$ and $F(3, -1)$, we can calculate the gradient of the line:

$$\text{gradient of } EF = \frac{y_F - y_E}{x_F - x_E}$$

$$= \frac{-1 - 2}{3 - 3}$$

$$= \frac{-3}{0}$$

There is no answer to this as division by 0 isn't possible. We say that the gradient of a vertical line is 'undefined'.
This makes some sense as a vertical line is so 'steep' that we cannot give it a value.

You will notice that the x-coordinate of **every** point on this vertical line is the same (3). So we can say that the equation of this line is $x = 3$. For every value of y, the value of x is always 3.

We can draw many other vertical lines on our Cartesian plane. Each vertical line will contain points that all have the same x-coordinate. In general, the equation for any vertical line will be $x = a$ where a is the constant number on the x-axis that the line passes through (the x-coordinate of each point on the line).

Calculate the gradient for some other vertical lines and you will find that it is always undefined as we cannot divide by 0.

- The gradient of a vertical line is **undefined**.
- The equation for a vertical line is $x = a$ where a is a constant number on the x-axis that the line passes through.
- Vertical lines are always **parallel to the y-axis**.

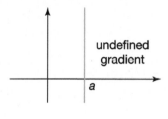

undefined gradient

a

Exercise 3

1 The points in each set lie on the same straight line. Write down the equation of each line.

a) $(-5, 2)$, $(-3, 2)$, $(0, 2)$, $(4, 2)$, $(7, 2)$

b) $(-1, -1)$, $(-5, -1)$, $(3, -1)$, $(10, -1)$, $(12, -1)$

c) $\left(-7, -\frac{1}{4}\right)$, $\left(-\frac{3}{4}, -\frac{1}{4}\right)$, $\left(0, -\frac{1}{4}\right)$, $\left(4, -\frac{1}{4}\right)$, $\left(8, -\frac{1}{4}\right)$

d) $(5, 14)$, $(5, 9)$, $(5, 1)$, $(5, -2)$, $(5, -5)$

e) $(-7, 3)$, $(-7, 2)$, $(-7, 1)$, $(-7, 0)$, $(-7, -1)$

f) $\left(\frac{2}{3}, -2\right)$, $\left(\frac{2}{3}, -3\right)$, $\left(\frac{2}{3}, -4\right)$, $\left(\frac{2}{3}, -5\right)$, $\left(\frac{2}{3}, -6\right)$

g) $(1, 0)$, $(2, 0)$, $(3, 0)$, $(4, 0)$, $(5, 0)$

h) $(9, 10)$, $(9, -10)$, $(9, 11)$, $(9, -11)$, $(9, 12)$

i) $\left(\frac{3}{4}, -\frac{1}{2}\right)$, $\left(\frac{3}{4}, \frac{1}{4}\right)$, $\left(\frac{3}{4}, -\frac{2}{3}\right)$, $\left(\frac{3}{4}, -\frac{1}{3}\right)$, $\left(\frac{3}{4}, 1\right)$

j) $(2, -6)$, $(4, -6)$, $(6, -6)$, $(8, -6)$, $(10, -6)$

2 Write down the equation for each of the lines shown on this Cartesian plane.

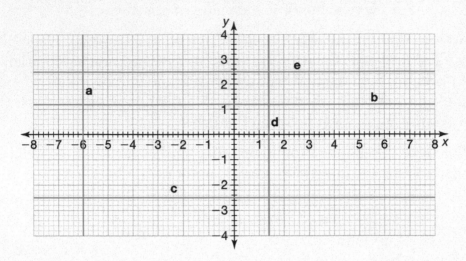

3 Write down the coordinates of three points on each of these lines.

a) $x = -2.3$ b) $y = 7.1$ c) $y = -\frac{3}{8}$ d) $x = 14$

4 Draw the graphs of these equations on the same Cartesian plane.

a) $x = -4.6$ b) $y = 1\frac{1}{2}$ c) $y = -3.8$ d) $x = 2.3$

D   The *y*-intercept

The point where a graph crosses the *y*-axis is called the **y-intercept**. Because this point is on the *y*-axis, its *x*-coordinate will always be 0. For this reason, we write the *y*-intercept as a single number, equal to the *y*-coordinate.

E   Equations in the form $y = mx + c$

For each of the following equations, we will:

1 draw the graph by working out pairs of values to plot as coordinates
2 calculate the gradient from two points on the line using the formula
3 find the *y*-intercept from the graph.

a) $y = 3x + 2$

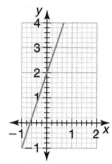

Gradient = 3
y-intercept = 2

b) $y = 4x + 3$

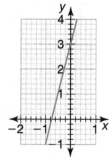

Gradient = 4
y-intercept = 3

c) $y = 2x - 3$

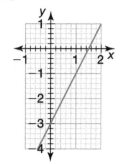

Gradient = 2
y-intercept = −3

d) $y = \frac{1}{2}x + 3$

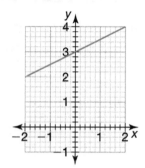

Gradient = $\frac{1}{2}$
y-intercept = 3

e) $y = 2x$

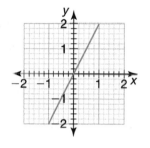

Gradient = 2
y-intercept = 0

f) $y = 3$

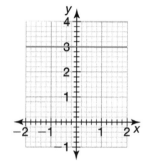

Gradient = 0
y-intercept = 3

g) $y = -2x + 4$

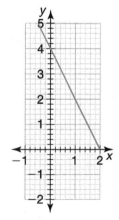

Gradient = −2
y-intercept = 4

h) $y = -\frac{1}{2}x + 4$

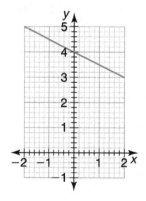

Gradient = $-\frac{1}{2}$
y-intercept = 4

We can summarise these facts in this table.

Equation	Gradient	y-intercept
$y = 3x + 2$	3	2
$y = 4x + 3$	4	3
$y = 2x - 3$	2	-3
$y = \frac{1}{2}x + 3$	$\frac{1}{2}$	3
$y = 2x$	2	0
$y = 3$	0	3
$y = -2x + 4$	-2	4
$y = -\frac{1}{2}x + 4$	$-\frac{1}{2}$	4

If we compare the **equation** of each line with the value of its **gradient** and **y-intercept**, we can see that the equation for each graph is in the form of:

$$y = (\text{gradient})x + (\text{y-intercept})$$

We can generalise this for **all** equations of straight lines:

> A line with an equation of the form $y = mx + c$ has a **gradient**, m, and **y-intercept**, c.

F Rearranging equations into the form $y = mx + c$

The general equation for a straight-line graph is $y = mx + c$.

When an equation is **in this form**, the gradient and the y-intercept are given by the values of m and c respectively.

The equation for a straight line can also be written in the form $px + qy = r$
But the gradient is **not** the coefficient of x when the equation is written like this. Also, the y-intercept of this line is not the constant number r.

To find the gradient and y-intercept of this line, we must **first rearrange** the equation.

In this case, we can rewrite it: $qy = -px + r$ and so $y = -\frac{p}{q}x + \frac{r}{q}$

So the gradient is $-\frac{p}{q}$ and the y-intercept is $\frac{r}{q}$.

Examples

a) Find the gradient and the y-intercept of the following straight lines.

   i) $4y + x = 2$

   Change the equation $4y + x = 2$ into the form $y = mx + c$.
   $4y + x = 2$
   So $4y = -x + 2$
   and $y = -\frac{1}{4}x + \frac{1}{2}$
   This gives us the gradient, $m = -\frac{1}{4}$
   and the y-intercept, $c = \frac{1}{2}$

   ii) $3y = -5x + 2$

   Change the equation $3y = -5x + 2$ into the form $y = mx + c$.
   $3y = -5x + 2$
   So $y = -\frac{5}{3}x + \frac{2}{3}$
   This gives us the gradient, $m = -1\frac{2}{3}$
   and the y-intercept, $c = \frac{2}{3}$

b) Write down the equations of these straight lines.

   i) Gradient is 3 and y-intercept is -2.

   Given: gradient, $m = 3$
   y-intercept, $c = -2$

   The equation of a straight line is $y = mx + c$

   So this equation is: $y = 3x - 2$

   ii) Gradient is $-\frac{1}{2}$ and y-intercept is 3.

   Given: gradient, $m = -\frac{1}{2}$
   y-intercept, $c = 3$

   The equation of a straight line is $y = mx + c$

   So this equation is: $y = -\frac{1}{2}x + 3$
   or $2y + x = 6$

c) For the straight line given in the graph, find its equation in the form $y = mx + c$.

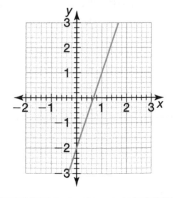

We have to find the gradient, m and the y-intercept, c.
Choose any two points on the line, say $(0, -2)$ and $(1, 1)$.

$$\text{Gradient, } m = \frac{y_2 - y_1}{x_2 - x_1}$$

$$= \frac{1 - (-2)}{1 - 0} = \frac{3}{1} = 3$$

From the graph, the y-intercept, $c = -2$

So the equation of this straight line is $y = 3x - 2$.

d) Work out the equation for each of the two parallel lines shown in this graph.

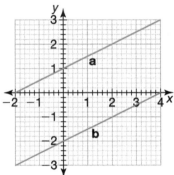

i) Two points on line **a** are $(0, 1)$ and $(2, 2)$.

$$\text{gradient} = \frac{y_2 - y_1}{x_2 - x_1} = \frac{2 - 1}{2 - 0} = \frac{1}{2}$$

y-intercept $= 1$
So the equation for line **a** is $y = \frac{1}{2}x + 1$.

ii) Two points on line **b** are $(0, -2)$ and $(2, -1)$.

$$\text{gradient} = \frac{y_2 - y_1}{x_2 - x_1} = \frac{-1 - (-2)}{2 - 0} = \frac{1}{2}$$

y-intercept $= -2$
So the equation for line **b** is $y = \frac{1}{2}x - 2$.

NOTE: The two lines in the example above are parallel. We find that both the lines have the same gradient $\left(\frac{1}{2}\right)$.
This is true for all parallel lines.

1 Any two lines that are parallel have the same gradient.
2 Two lines that have the same gradient are parallel.

Example

The equation of a straight line is $8x + 4y = 2$.
Write down the equation of one other line that is **parallel** to this line.

Write the equation in the form $y = mx + c$.

$$8x + 4y = 2$$
$$4y = -8x + 2$$
$$y = -2x + \frac{1}{2}$$

So the gradient of this line is -2.
We know that any line that is parallel to this line will also have a gradient of -2.
Parallel lines cannot pass through the same point, so a line that is different from the given line will have a different y-intercept.
We are not given any more information, so we can choose any y-intercept, for example -3.
This will give the equation: $y = -2x - 3$.

Can you write the equation of another line that is parallel to these two lines?

G The x-intercept

We call the point where a straight-line graph crosses the y-axis the **y-intercept**. In the same way, we call the point where a straight-line graph crosses the x-axis the **x-intercept**. At all points on the x-axis, $y = 0$. So we can work out the x-intercept from the equation of the line by substituting the value of $y = 0$ and solving for x.

 Exercise 4

1 For each of these straight lines, write down
 i) the gradient and ii) the y-intercept.

a) $5y + 4x + 20$ b) $2y = x + 1$ c) $4x + y = 3$ d) $2x + 5y = 10$
e) $2y - 5 = x$ f) $2x - 7y = 14$ g) $6x + 3y = 8$ h) $-2x + 3y = -21$
i) $x + 5y = 2$ j) $2x - 5y = 3$ k) $3x - 5y = 5$ l) $4x - 2y = 7$
m) $6x - y = 4$ n) $-2x + 7y = 0$ o) $x + 4 = 4y$ p) $4x - 6y + 3 = 0$

2 **i)** Write down the equation of the straight line that has the gradient and *y*-intercept given.

ii) Write each equation using only positive whole-number coefficients and constants.

	Gradient	*y*-intercept
a)	6	−4
b)	$\frac{1}{2}$	3
c)	$-\frac{2}{3}$	$\frac{3}{7}$
d)	−2.5	0
e)	0	−13
f)	$\frac{3}{4}$	$-1\frac{1}{3}$

3 Work out the equation of each of the lines drawn on this graph.

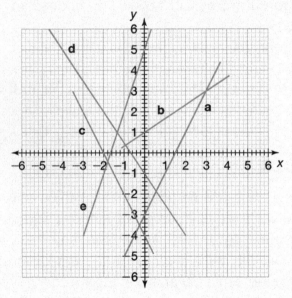

4 Write down the equation of the line that is parallel to the given line and cuts the *y*-axis at the given point.

	Point	Equation
a)	$(0, 2)$	$y = 2x + 6$
b)	$\left(0. -\frac{1}{2}\right)$	$y = 2x + 4$
c)	$(0, -4)$	$y = -3x + 7$
d)	$(0, -11)$	$2y = x - 1$
e)	$(0, -1)$	$y + 4 = 9x$
f)	$\left(0, \frac{1}{3}\right)$	$2x + 3y = 6$
g)	$(0, -0.8)$	$4x + y = -2$
h)	$(0, -7)$	$2x + 3y = 1$

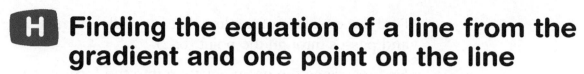

H Finding the equation of a line from the gradient and one point on the line

We can use the gradient of a straight line and **one** point on that line to work out the *y*-intercept, and, from this, also find the equation of the line.

Examples

a) A straight line has a gradient of 2 and passes through the point *A*(1, 3). Find the *y*-intercept and the equation of this line.

The gradient of the line is 2, so we know that $y = 2x + c$, where *c* is the *y*-intercept.

This line passes through point *A*, so we also know that, when $x = 1$, $y = 3$, these values make the equation true.

Substituting $x = 1$ and $y = 3$ into $y = 2x + c$, we have
$$3 = (2 \times 1) + c,$$
$$3 = 2 + c$$
$$c = 1$$

With a *y*-intercept of 1, the full equation of this line is $y = 2x + 1$.

b) Find the equation of the line that passes through (1, 2) and has a gradient of −3.

From the gradient, $m = -3$, we know the equation of the line is $y = -3x + c$.

Substituting the values of *x* and *y* from the point (1, 2) into this equation, we have

$$2 = (-3 \times 1) + c,$$
$$2 = -3 + c$$
$$c = 2 + 3 = 5$$

so the equation of the line is $y = -3x + 5$.

c) Find the equation of the line that passes through (−1, 2) and has an undefined gradient.

As the line has an undefined gradient, it is a vertical line whose equation is of the form $x = a$ where *a* is a constant. Since it passes through (−1, 2), its equation is $x = -1$.

d) Find the equation of the line that is parallel to $3y + 2x + 6 = 0$ and passes through the point $(6, 2)$.

First rearrange the given equation into the form $y = mx + c$.

$$3y + 2x + 6 = 0$$
$$3y = -2x - 6$$
$$y = -\tfrac{2}{3}x - 2$$

So the gradient of the given line is $m = -\tfrac{2}{3}$

and the equation of the parallel line is $y = -\tfrac{2}{3}x + c$

Now substitute the point $(6, 2)$ into this new equation to give:

$$2 = -\left(\tfrac{2}{3} \times 6\right) + c$$
$$2 = -4 + c$$
$$c = 6$$

The equation of the new line is $y = -\tfrac{2}{3}x + 6$.

1 Finding the equation of a line passing through two given points

If we are given only **two** points on a straight line, we already know how to use these to work out the gradient. With the gradient and only one of the points, we can work out the y-intercept for the line, and, from this, the equation for the line.

Example

Find the equation of the line passing through $A(1, 3)$ and $B(2, 5)$.

The gradient of AB, $\quad m = \dfrac{y_2 - y_1}{x_2 - x_1}$

$$= \dfrac{5 - 3}{2 - 1} = 2$$

So the equation of the line is $y = 2x + c$.

Substituting **one** of the points (either one will do) into this equation, we have

$$3 = (2 \times 1) + c,$$
$$3 = 2 + c$$
$$c = 1$$

With a y-intercept of 1, the full equation of this line is $y = 2x + 1$.

Exercise 5

1 Work out the equation for each straight line, given the gradient and one point on the line.

	Gradient	Point
a)	2	(3, 2)
b)	−5	(−1, 3)
c)	−1	(5, −2)
d)	$-\frac{2}{3}$	(3, 4)
e)	$-\frac{1}{4}$	(2, 1)
f)	−2	(0, 0)
g)	0	$\left(\frac{1}{2}, 2\right)$
h)	undefined	(5, −6)

2 Write down the equation of the line that is parallel to the given line and passes through the given point.

	Point	Equation
a)	(4, 3)	$x + 2y = 3$
b)	(5, 2)	$x - 2y + 3 = 0$
c)	(1, −4)	$x + 4 = 0$
d)	(1, 3)	$y = 2x + 1$
e)	(−4, −1)	$4y + x = -8$
f)	(5, −6)	$2x + y = -11$

3 Find the equation of the line that passes through each of these pairs of points.

a) (0, 5) and (2, 0) b) (1, 5) and (2, 7)

c) (2, −2) and (7, 8) d) (2, −3) and (4, −2)

e) $\left(\frac{1}{2}, 1\right)$ and $\left(\frac{3}{4}, \frac{1}{2}\right)$ f) (−3, −5) and (−1, 3)

g) (4, 1) and (−4, 7) h) (−1, −4) and (2, 0)

Unit 2 Linear inequalities

In Coursebook 1 we learned about numbers and how to show them in order by using a number line. On a number line, we show the **bigger** numbers on the right and the **smaller** numbers on the left. This means that any number on the number line is bigger than every other number to its left and smaller than every other number to its right. We use the symbol '>' to stand for 'is bigger than' (or 'is greater than') and we use the symbol '<' to stand for 'is smaller than' (or 'is less than').

Look at this number line.

$$\xleftarrow{\hspace{0.5cm}} \overset{\displaystyle -5\ -4\ -3\ -2\ -1\ \ 0\ \ 1\ \ 2\ \ 3\ \ 4\ \ 5}{|\ |\ |\ |\ |\ |\ |\ |\ |\ |\ |} \xrightarrow{\hspace{0.5cm}}$$

4 is to the right of 1 and so we know that 4 is bigger than 1. We write this as $4 > 1$.

-5 is to the left of -2 and so we know that -5 is smaller than -2. We write this as $-5 < -2$.

Remember: Even though we usually show only the whole numbers on a number line, there are many other real numbers between each two whole numbers.

If we look at any two real numbers, a and b, on a number line, then we can say:

either	$a > b$
or	$a = b$
or	$a < b$,

but only **one** of these can be true.

Equations and inequalities

In Coursebook 2 we learned about linear equations and how to solve them.

A **linear equation** is usually true for one fixed and definite value of the unknown variable.

In Unit 3 of this book we will learn about quadratic equations. A **quadratic equation** is true for two fixed and definite values of the unknown variable.

When a number is bigger than or smaller than another number, the mathematical statement we write is called an **inequality**.
Examples of inequalities are: $x > 2$ or $y < -3$.

We see that an inequality is not true for just one or two definite values of the variable. An inequality is true for **many** different values (an infinite number) of the variable at the same time. We say that the unknown variable in an inequality is true for a whole **range** of values.

Look at the examples above.

For $x > 2$, x can be **any** number, as long as it is bigger than 2 $\left(2\frac{1}{2}, 3.35, 5.7897, 8.\dot{6}, \ldots\right)$.

For $y < -3$, y can be **any** number, as long as it is smaller than -3 $\left(-3\frac{1}{4}, -5.46, -7.92347, \ldots\right)$.

To make things a bit easier, we often use only the **integer values** of the variable (as shown on the number line) because it is impossible to list all the real number values.

We can easily find examples of inequalities used in daily life. For example, children under the age of 18 may not vote in an election. We could write this as:

People whose age is less than 18 years (age < 18) may not vote.

or People whose age is greater than 18 years (age > 18) may vote.

You will probably ask: 'What about people whose age is equal to 18 years?' We will see later in this unit how we can combine an equality with inequalities like this.

Can you think of other examples?

Exercise 1

1 Copy the statements and write '>', '=' or '<' in each space to make the statement true.

a) $6 \,\square\, 9$

b) $-3 \,\square\, 1$

c) $0 \,\square\, -4$

d) $-0.6 \,\square\, -0.5$

e) $3 \,\square\, \sqrt{3}$

f) $3^2 \,\square\, (-3)^2$

g) $(-3)^2 \,\square\, (-4)^2$

h) $(-3)^3 \,\square\, (-4)^3$

i) $3^2 \,\square\, -3^2$

j) $\frac{1}{3} \,\square\, \frac{1}{4}$

k) $-\frac{1}{3} \,\square\, -\frac{1}{4}$

l) $\left(\frac{1}{3}\right)^2 \,\square\, \left(\frac{1}{4}\right)^2$

m) $\left(-\frac{1}{3}\right)^2 \,\square\, \left(-\frac{1}{4}\right)^2$

n) $\left(-\frac{1}{3}\right)^3 \,\square\, \left(-\frac{1}{4}\right)^3$

o) $(0.6)^2 \,\square\, 0.6$

p) $(0.6)^2 \,\square\, 0.36$

q) $\sqrt{0.6} \,\square\, 0.6$

r) $3 + y \,\square\, 4 + y$

s) $k - 10 \,\square\, k - 8$

t) $12 - 3t \,\square\, 4 - 3t$

B Properties of inequalities

Many of the properties of inequalities are the same as those of equations, but there is one important difference with inequalities. All the properties listed below are true for any inequalities, whether they involve > or <. The examples given use only the inequality >. Choose your own examples to make sure they are true for < as well.

1 Consider any three numbers, p, q and r. If $p > q$ and $q > r$, then $p > r$

(e.g. using the numbers 8, 4 and -3: $8 > 4$ and $4 > -3$, so $8 > -3$).

2 We can add or subtract any number from both sides of an inequality and this will not change the inequality sign.

a) If x is a **positive** number (e.g. 3) and p and q are any two numbers so that $p > q$ (e.g. $7 > 1$)

then $p + x > q + x$ (e.g. $7 + 3 > 1 + 3$)

and $p - x > q - x$ (e.g. $7 - 3 > 1 - 3$)

b) If y is a **negative** number (e.g. -2) and p and q are any two numbers so that $p > q$ (e.g. $7 > 1$)

then $p + y > q + y$ (e.g. $7 - 2 > 1 - 2$)

and $p - y > q - y$ (e.g. $7 + 2 > 1 + 2$)

3 We can multiply or divide both sides of an inequality by the same **positive** number and this will not change the inequality sign.

If x is a **positive** number (e.g. 3) and p and q are any two numbers so that $p > q$ (e.g. 7 > 1)

then $\qquad px > qx \qquad$ (e.g. 7 × 3 > 1 × 3)

and $\qquad \dfrac{p}{x} > \dfrac{q}{x} \qquad$ (e.g. 7 ÷ 3 > 1 ÷ 3)

4 Property **3** is **not** true when we multiply or divide by a **negative** number. We must **change the inequality sign** if we multiply or divide by a **negative** number.

If y is a **negative** number (e.g. −2) and p and q are any two numbers so that $p > q$ (e.g. 7 > 1)

then $\qquad py < qy \qquad$ (e.g. 7 × (−2) < 1 × (−2))

and $\qquad \dfrac{p}{x} < \dfrac{q}{x} \qquad$ (e.g. 7 ÷ (−2) < 1 ÷ (−2))

Exercise 2

1 Copy the statements and write '>', '=' or '<' in each space to make the statement true.

a) If $x < 25$ and $25 < y$, then $x \,\square\, y$ b) If $x > y$ and $y > -5$, then $x \,\square\, -5$

c) If $x + 4 = y$ then $x \,\square\, y$ d) If $x = 2 - y$ and $y > 0$, then $2 \,\square\, x$

e) If $y > x$, then $6x \,\square\, 6y$ f) If $x < y$, then $\dfrac{x}{12} \,\square\, \dfrac{y}{12}$

g) If $y < x$, then $\dfrac{x}{-4} \,\square\, \dfrac{y}{-4}$ h) If $x > y$, then $(-3)x \,\square\, (-3)y$

i) If $x < 0$, then $(-8)x \,\square\, 0$ j) If $y > 0$, then $\dfrac{y}{-50} \,\square\, 0$

k) If $x - 6 = y$, then $y \,\square\, x$ l) If $x < y$ and $y < 0$, the $x \,\square\, 0$

C Solving inequalities

As with equations, to **solve** an inequality means to find the values for the unknown variable which make the inequality true. We can use the properties of inequalities to help us solve them in almost the same way as we learned to solve equations. Again, the aim is to end up with **one variable** on one side of the inequality and a **number** on the other side.

Examples

Solve these inequalities.

a) $x + 4 > 11$

Subtract 4 from both sides:
$x + 4 - 4 > 11 - 4$
$x > 7$

b) $5x - 3 < 27$

Add 3 to both sides:
$5x - 3 + 3 < 27 + 3$
$5x < 30$
Divide both sides by 5:
$x < 6$

c) $7 - x > 2$

Subtract 7 from both sides:
$7 - x - 7 > 2 - 7$
$-x > -5$
Multiply both sides by -1 (remember to change the inequality sign):
$x < 5$

Sometimes the expressions on each side of the inequality may be more complex.

Examples

Solve these inequalities.

a) $2(3x - 4) < 3(x + 1) - 5$

Expand the brackets:
$6x - 8 < 3x + 3 - 5$
Add 8 to both sides:
$6x - 8 + 8 < 3x - 2 + 8$
$6x < 3x + 6$
Subtract $3x$ from both sides:
$6x - 3x < 3x + 6 - 3x$
$3x < 6$
Divide both sides by 3:
$x < 2$

b) $\frac{1}{4}(11 - x) > 2 - x$

Multiply both sides by 4 to remove fractions:
$4 \times \frac{1}{4}(11 - x) > 4(2 - x)$
$11 - x > 8 - 4x$
Add $4x$ to both sides:
$11 - x + 4x > 8 - 4x + 4x$
$11 + 3x > 8$
Subtract 11 from both sides:
$11 + 3x - 11 > 8 - 11$
$3x > -3$
$x > -1$

D Number lines

One way that we can show an inequality is on a number line:

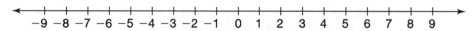

- As you move to the right, numbers get bigger.
- As you move to the left, numbers get smaller.
- For an inequality, the starting number is **not included** (only the values bigger than or smaller than this number), so we show the start of the inequality with an **open** circle.
- An inequality has **no end**, so we finish with an **arrowhead**.

So the solutions to the examples on page 26 look like this on a number line.

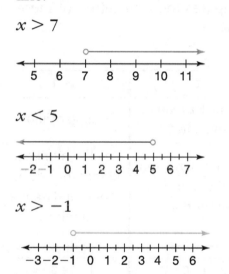

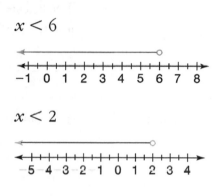

E Combining equations and inequalities

So far, we have studied equations and inequalities separately. Sometimes we can have solutions that are combinations of an inequality and an equation. For example, the solution could be that x is any number smaller than **or equal** to 4. The solutions $x = 4$ or $x < 4$ are both possible. This time, the number 4 is included in the list of numbers that make the inequality true.

The symbol we use for 'is smaller than or equal to' is $\leq$, so we would write the solution as $x \leq 4$.

This solution is shown on a number line by using a closed dot at the beginning of the arrow instead of the open circle we used for pure inequalities.

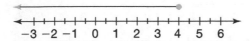

We can also combine the inequality 'is bigger than' with an equation to give the new inequality 'is bigger than or equal to', and we use the symbol $\geqslant$. So, for example, the inequality $x \geqslant -2$ would be shown on a number line as:

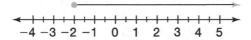

The table below compares the four different inequalities that we have learned: $<$, $\leqslant$, $>$ and $\geqslant$.

	< 1	$\leqslant 1$	> 1	$\geqslant 1$
We read it as …	'is smaller than 1'	'is smaller than or equal to 1'	'is bigger than 1'	'is bigger than or equal to 1'
It means …	all the numbers smaller than 1, but not including the number 1	the number 1 and all the numbers smaller than 1	all the numbers bigger than 1, but not including the number 1	the number 1 and all the numbers bigger than 1
On a number line we draw …	$\xleftarrow{\quad\quad}\!\!\!\!\!\!\overset{\circ}{}$ −2 −1 0 1	$\xleftarrow{\quad\quad}\!\!\!\!\!\!\overset{\bullet}{}$ −2 −1 0 1	$\overset{\circ}{}\!\!\!\!\!\!\xrightarrow{\quad\quad}$ 1 2 3 4	$\overset{\bullet}{}\!\!\!\!\!\!\xrightarrow{\quad\quad}$ 1 2 3 4

Examples

Solve these inequalities and show the solution on a number line.

a) $2(x - 5) \geqslant 2 - x$

$2x - 10 \geqslant 2 - x$
Add $x + 10$ to both sides:
$2x - 10 + x + 10 \geqslant 2 - x + x + 10$
$\qquad\qquad\qquad 3x \geqslant 12$
Divide both sides by 3:
$\qquad\qquad\qquad x \geqslant 4$

b) $\dfrac{5x}{6} - 1 \geqslant \dfrac{1}{2}(4x - 9)$

Multiply both sides by LCM of 2 and 6:

$$6 \times \dfrac{5x}{6} - 6 \geqslant 6 \times \dfrac{1}{2}(4x - 9)$$

$$5x - 6 \geqslant 3(4x - 9)$$
$$5x - 6 \geqslant 12x - 27$$

Add $27 - 5x$ to both sides:

$$5x - 6 + 27 - 5x \geqslant 12x - 27 + 27 - 5x$$
$$21 \geqslant 7x$$

Divide both sides by 7:

$$x \leqslant 3$$

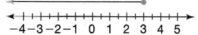

$$-4\,-3\,-2\,-1\ \ 0\ \ 1\ \ 2\ \ 3\ \ 4\ \ 5$$

Exercise 3

1 Solve each of these inequalities and show the solution on a number line.

a) $4x - 7 > 17$

b) $10 - 8x > 26$

c) $-6x - 1 \leqslant 23$

d) $32 - 15x < 2$

e) $-26 < 4 - 5x$

f) $26 \leqslant -7x - 2$

g) $10x + 18 \geqslant -72$

h) $12 > -14x - 2$

i) $58 - x \geqslant 20$

j) $3x - 8 > 10$

k) $-2x + 7 \leqslant 37$

l) $\dfrac{x}{5} - 9 > 3$

m) $7 - \dfrac{x}{10} \geqslant 12$

n) $\dfrac{2}{5}x - 5 \geqslant 3$

o) $-12 \geqslant 8 - \dfrac{4}{3}x$

p) $13 - \dfrac{3}{4}x > 13$

2 Solve each of these inequalities and show the solution on a number line.

a) $8 + 2x \leqslant 6x - 20$

b) $15 - 4x \geqslant 2x + 23$

c) $9x - 99 \geqslant 18x$

d) $2 - 13x < 8 - 11x$

e) $28 < 4(5 - 2x)$

f) $8 - 2x < 5 - 5x$

g) $3x - 10 \leqslant 7(2 + x)$

h) $3 - 2x > x + 15$

i) $-4(2x - 6) < x + 6$

j) $3(x - 1) < 2(3x - 4)$

k) $2(x + 1) > 5(4 - x)$

l) $6(3x + 4) > 10(2x - 1)$

m) $7(3x - 2) > 9(x + 3)$

n) $4(1 - 3x) - 14 > 4(2x + 3) - 9x$

3 Solve each of these inequalities.

a) $\dfrac{4 - x}{3} > \dfrac{1}{-2}$

b) $\dfrac{x + 1}{4} - \dfrac{1}{12} \leqslant \dfrac{x}{3}$

c) $\dfrac{2(4x - 3)}{5} \leqslant -6$

d) $\dfrac{3 - 4x}{5} \geqslant 2\dfrac{1}{2} - 6x$

e) $\dfrac{2x}{3} - \dfrac{x}{6} < -2$

f) $\dfrac{3x - 1}{10} \geqslant \dfrac{x - 1}{4} + \dfrac{3}{10}$

g) $\dfrac{x + 2}{3} - \dfrac{x + 1}{4} \geqslant 2$

h) $\dfrac{x + 2}{2} + \dfrac{x - 1}{5} > \dfrac{1}{10}$

i) $\dfrac{x - 1}{5} + \dfrac{3x + 2}{5} \leqslant \dfrac{21 - 3x}{4}$

j) $0.2(x + 1) < 0.5(2x - 1)$

F More than one linear inequality

Sometimes we find that a variable must satisfy more than one inequality at the same time. The linear inequalities are usually joined by the word 'and'. For example, if $x \geqslant -2$ and $x < 7$, then we can say that x can be any value that is from -2 (including the number -2) up to 7 (but not including the number 7). We say that x can be any value in this **range** of numbers.

This can also be written in a short way to combine both inequalities. We usually begin with the smallest value:

$$-2 \leqslant x < 7$$

Make sure the inequalities read the right way round from the variable.

It is also convenient sometimes to write down all the **integer values** of the variable that are in the range of the inequalities. The integer values that make these two inequalities true are:

$-2, -1, 0, 1, 2, 3, 4, 5, 6$ (remember that the number 7 is not included in this range of values)

Now we are going to solve combined inequalities like this so that we have just **one variable** in a range between **two numbers**.

Examples

a) Find the values of x so that $3x - 2 > -8$ and $4x - 2 \leqslant 10$ and show the solution on a number line.

Solve each of the inequalities separately.

$$
\begin{array}{ll}
3x - 2 > -8 & 4x - 2 \leqslant 10 \\
\quad\;\; 3x > -6 & \quad\;\; 4x \leqslant 12 \\
\quad\;\;\; x > -3 & \quad\;\;\; x \leqslant 3
\end{array}
$$

So the values of x must be bigger than -3 and also smaller than or equal to 3, written: $-3 < x \leqslant 3$.

On the number line this looks as follows:

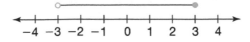

b) Find the **integer** values of x so that $5x - 3 < 27$ and $6x + 1 \geqslant 4x - 2$.

Solve each of the inequalities separately.

$$5x - 3 < 27 \qquad\qquad 6x + 1 \geqslant 4x - 2$$
$$5x < 30 \qquad\qquad\qquad 2x \geqslant -3$$
$$x < 6 \qquad\qquad\qquad\quad x \geqslant -1\tfrac{1}{2}$$

So the values of x are in the range $-1\tfrac{1}{2} \leqslant x < 6$. If we draw the solution on a number line, it is easier to see the integer values that are in this range:

Even though the number $-1\tfrac{1}{2}$ is included in the solution range, it is **not** an integer. So the first integer in the solution range is -1.

All the integers in the solution range are $-1, 0, 1, 2, 3, 4, 5$ (6 is not included).

c) Find the solution for the inequalities $4x + 5 > 12 - 3x$ and $4x + 2 \leqslant 2x - 10$

Solve each of the inequalities separately.

$$4x + 5 > 12 - 3x \qquad 4x + 2 \leqslant 2x - 10$$
$$7x > 7 \qquad\qquad\qquad 2x \leqslant -12$$
$$x > 1 \qquad\qquad\qquad\quad x \leqslant -6$$

So the values of x must be smaller than -6 and bigger than 1. There is no value of x that satisfies both of these inequalities at the same time. Look at the solutions on a number line:

You can see that the two groups of numbers in the solutions do not overlap, and so we say there are **no solutions** to these two inequalities.

d) If $5 \leqslant x \leqslant 7\tfrac{1}{2}$ and $-5\tfrac{1}{4} \leqslant y \leqslant -2\tfrac{1}{3}$, find the following values.

i) The smallest possible rational value of $x - y$

$x - y$ is smallest when x is smallest and y is biggest,

i.e. when $x = 5$ and $y = -2\tfrac{1}{3}$

So $\quad x - y = 5 - (-2\tfrac{1}{3})$

$\qquad\qquad = 7\tfrac{1}{3}$

The smallest possible rational value of $x - y$ is $7\tfrac{1}{3}$.

ii) The biggest possible integer value of $x - y$

$x - y$ is biggest when x is biggest and y is smallest,

i.e. when $x = 7\frac{1}{2}$ and $y = -5\frac{1}{4}$

So $\quad x - y = 7\frac{1}{2} - (-5\frac{1}{4})$

$\qquad\qquad = 12\frac{3}{4}$

The biggest possible integer value of $x - y$ is 12.

G Double inequalities

If two inequalities are combined, we have learned that we can write their solutions in a combined way as well.
For example, the solutions $x < 4$ and $x \geqslant 0$ can be written $0 \leqslant x < 4$.

Sometimes we find that the inequalities themselves are written in this combined form, such as $-1 < 5x + 4 \leqslant 9$. These are called double inequalities as there are two inequality signs.

To solve a double inequality, we can use the same properties as before so that we end up with a single variable in a range between two numbers as the solution.

NOTE: Whatever you do to one part of the double inequality, you must also do to **all the other parts**. It is also possible to split the double inequality into two single inequalities and solve these as we have done before.

Example

Find the integer values of x so that $-1 < 5x + 4 \leqslant 9$.

Method 1: $\quad -1 < 5x + 4 \leqslant 9$
Subtract 4 from **every** part of the inequality:
$-5 < 5x \leqslant 5$
Divide each part by 5:
$-1 < x \leqslant 1$

Method 2: Solve each part as separate inequalities.

$-1 < 5x + 4$	$5x + 4 \leqslant 9$
Subtract 4 from both sides:	Subtract 4 from both sides:
$-5 < 5x$	$5x \leqslant 5$
Divide both sides by 5:	Divide both sides by 5:
$-1 < x$	$x \leqslant 1$

The combined solution range is $-1 < x \leqslant 1$.

Both methods give us the same range of solution values for x. The **integer** values for the solutions are $-1, 0, 1$.

Exercise 4

1 Find the solutions for each pair of inequalities and show the solutions on a number line.
If there are no solutions, then say so.

a) $x + 5 > 4$ and $x - 2 < 2$

b) $2x > 5$ and $3x + 2 < 95$

c) $5x - 1 < 4$ and $3x + 5 \geqslant x + 1$

d) $-3x > 12$ and $5x \geqslant 10$

e) $x - 1 < 10$ and $4x + 1 > 7$

f) $2x + 5 < 15$ and $3x - 2 > -6$

g) $-4x + 9 > 1$ and $7x - 13 \leqslant -6$

h) $2x - 3 \leqslant 5$ and $7 - 4x \leqslant -3$

i) $\frac{1}{2}x - 4 > \frac{1}{3}x$ and $\frac{1}{6}x + 1 > \frac{1}{8}x + 3$

j) $32 \leqslant 3x + 20$ and $17 > 1 - 8x$

2 Write down the integer values of x that are true for each of these double inequalities.

a) $-4 \leqslant 5x + 6 < 11$

b) $-3 \leqslant 4 - 2x \leqslant 12$

c) $0 < 2x - 8 < 3$

d) $3 < 5x - 1 \leqslant 29$

e) $x + 2 \geqslant 1 - 3x > x - 11$

f) $2(1 - x) > x - 1 \geqslant \frac{1}{2}(x - 7)$

g) $3x + 2 < 19 < 5x - 4$

h) $3x - 5 < x + 1 \leqslant 2x + 1$

i) $2x - 3 < x + 3 < 3x + 8$

j) $2 - x < 2x + 3 \leqslant x + 6$

k) $3x - 3 \leqslant 5x + 9 \leqslant x + 35$

l) $3x - 7\frac{1}{2} > 4x - 16\frac{1}{2} \geqslant x + 2\frac{1}{4}$

3 For each pair of inequalities, find:

i) the smallest possible rational value of $x - y$

ii) the biggest possible integer value of of $x - y$.

a) $-8\frac{1}{3} \leqslant x \leqslant -1.5$ and $-10\frac{1}{4} \leqslant y \leqslant -2$

b) $2 \leqslant x \leqslant 6\frac{1}{3}$ and $1\frac{2}{3} \leqslant y \leqslant 10\frac{3}{8}$

c) $-3\frac{3}{4} \leqslant x \leqslant 4\frac{5}{8}$ and $-4\frac{1}{4} \leqslant y \leqslant -1\frac{1}{8}$

d) $-4.6 \leqslant x \leqslant -\frac{1}{2}$ and $-1\frac{3}{4} \leqslant y \leqslant 7\frac{4}{5}$

Quadratic equations

Key vocabulary

complete the square	perfect square	solution set
completed square form	quadratic equation	zero product theorem
factorise	quadratic formula	

We have already learned how to solve equations where the highest degree of the variable is 1. These equations are called **linear equations**.

For example, $5x - 3 = 0$ gives $x = \frac{3}{5}$ when we solve the equation.

Now we are going to look at equations where the highest degree of the variable is 2. These equations are called quadratic equations.

> A **quadratic equation** is of the form
> $$ax^2 + bx + c = 0$$
> where a, b and c are constants (numbers) and $a \neq 0$.

For example, $2x^2 - 6x + 4 = 0$ gives $x = 1$ or $x = 2$ when we solve the equation.

How do we get these **two** answers for the value of x that are both true?

A Solving a quadratic equation by factorising

Activity

Think of two numbers.
Multiply these two numbers together.
The answer is zero.

1 Write down any two numbers that **could** be the numbers you thought of.
2 Now write down four more pairs of numbers that **could** be your numbers.
3 What can you say about your numbers?

You can see that **at least one** of the numbers must be zero so that the product of the two numbers is also zero. We use this fact to solve quadratic equations. It is called the **zero product theorem**.

> If the product of two factors is equal to zero, then **one or both** of the factors must be zero.

$$\text{If } a \times b = 0, \text{ then } \begin{cases} a = 0 \\ \text{or } b = 0 \\ \text{or both } a = 0 \text{ and } b = 0 \end{cases}$$

Examples

Find the values of x that make these equations true (i.e. solve for x).

a) $x(x - 7) = 0$

From the zero product theorem, we know that *either* $x = 0$ *or* $x - 7 = 0$.
So the values of x that make the original equation true are:
$x = 0$ *or* $x = 7$
(Check by substituting these values into the equation.)

b) $(x + 4)(x - 5) = 0$

From the zero product theorem, we know that *either*
$x + 4 = 0$ *or* $x - 5 = 0$.
So the values of x that make the original equation true are:
$x = -4$ *or* $x = 5$
(Check by substituting these values inot the equation.)

NOTE: The values of the variable that make an equation true are called the **solution set** for that equation. We write this solution set in 'curly' brackets like this: $\{-, -\}$.

So the solution sets for the example above would be:
a) $\{0, 7\}$ and **b)** $\{-4, 5\}$

The simplest way to solve quadratic equations is to factorise the quadratic expression as we learned in Coursebook 2 Unit 9.
This gives us factors multiplied together that equal zero, and we can use the zero product theorem again.

Examples

Solve these equations by factorising the quadratic expressions.

a) $x^2 - 3x - 18 = 0$

$(x - 6)(x + 3) = 0$
either $\quad x - 6 = 0 \quad$ or $\quad x + 3 = 0$
i.e. $\qquad\quad x = 6 \quad$ or $\qquad x = -3$
So the solution set for this equation is:
$\{6, -3\}$

b) $x^2 - 5x + 6 = 0$

$(x - 2)(x - 3) = 0$
either $\quad x - 2 = 0 \quad$ or $\quad x - 3 = 0$
i.e. $\qquad\quad x = 2 \quad$ or $\qquad x = 3$
So the solution set for this equation is:
$\{2, 3\}$

c) $6x^2 + 5x - 4 = 0$

$(2x - 1)(3x + 4) = 0$
either $\qquad 2x - 1 = 0 \quad$ or $\quad 3x + 4 = 0$
i.e. $\qquad\qquad x = \frac{1}{2} \quad$ or $\qquad x = -\frac{4}{3}$
So the solution set for this equation is:
$\left\{\frac{1}{2}, -\frac{4}{3}\right\}$

d) $a^2 = 7a$

We must write the equation so that it equals 0
So $\quad a^2 - 7a = 0$
$\qquad a(a - 7) = 0$
$\qquad$ either $\quad a = 0 \quad$ or $\quad a - 7 = 0$
$\qquad\quad$ i.e. $\quad a = 0 \quad$ or $\qquad a = 7$
Solution set: $\quad \{0, 7\}$

e) $y^2 = 9$

So $\quad y^2 - 9 = 0 \quad$ (this is a difference of perfect squares)
$(y + 3)(y - 3) = 0$
either $\quad y + 3 = 0 \qquad$ or $\quad y - 3 = 0$
i.e. $\qquad\quad y = -3 \quad$ or $\qquad y = 3$
Solution set: $\quad \{3, -3\}$

f) $a^2 + 4a = 5$

So $\quad a^2 + 4a - 5 = 0$
$(a - 1)(a + 5) = 0$
either $\quad a - 1 = 0 \quad$ or $\quad a + 5 = 0$
i.e. $\qquad\quad a = 1 \quad$ or $\qquad a = -5$
Solution set: $\quad \{1, -5\}$

To solve a quadratic equation by factorising

1 Write the equation so that every term is on the left-hand side (LHS), and only '0' is on the right-hand side (RHS).
2 Arrange the terms on the LHS in descending order of the powers of the variable.
3 Factorise the quadratic expression.
4 Use the zero product theorem to make each factor equal to zero. Then solve these equations to find the values for the unknown variable letter.
5 Check each of the values in this solution set in the original equation.

What happens if we are given the factors, but their product is not equal to zero?

Example

Solve the following equation: $(x - 1)(x - 3) = 8$

NOTE: It is **not** correct to say that *either* $x - 1 = 8$ *or* $x - 3 = 8$. We can only use the zero product theorem when the product of factors equals 0.

So the first thing to do is to multiply the brackets and then take the 8 to the LHS and include it in the quadratic expression **before** factorising. Then the product of factors will equal 0.

$$(x - 1)(x - 3) = 8$$
$$x^2 - 4x + 3 = 8$$
$$x^2 - 4x + 3 - 8 = 0$$
$$x^2 - 4x - 5 = 0$$

Now we can factorise the expression and use the zero product theorem because the new expression does equal 0.
$$(x - 5)(x + 1) = 0$$
either $x - 5 = 0$ *or* $x + 1 = 0$
$\qquad\qquad x = 5$ *or* $\qquad x = -1$
Solution set: $\{5, -1\}$

Exercise 1

1 Solve these equations.

a) $(x + 4)(x - 1) = 0$ b) $(x - 2)(x - 6) = 0$ c) $(x + 1)(x - 3) = 0$
d) $(2a - 3)(a + 4) = 0$ e) $m(m - 5) = 0$ f) $6x(3x + 2) = 0$

2 Solve these equations by factorising the quadratic expression.

a) $x^2 - 11x + 24 = 0$ b) $y^2 + 7y + 12 = 0$ c) $m^2 + 9m + 20 = 0$
d) $a^2 + a - 12 = 0$ e) $n^2 - 8n + 15 = 0$ f) $z^2 - 9z + 18 = 0$
g) $n^2 - 18n + 17 = 0$ h) $c^2 + 15c + 56 = 0$ i) $x^2 - 13x + 40 = 0$
j) $y^2 - 7y - 60 = 0$ k) $a^2 - 7a - 18 = 0$ l) $p^2 - p - 72 = 0$

3 Solve these equations by factorising the quadratic expression.

a) $x^2 - 5x = 0$ b) $3a^2 - 4a = 0$ c) $b^2 + 3b = 0$
d) $3d = 81d^2$ e) $g^2 = 6g$ f) $7x^3 + 21x^2 = 0$

4 Solve these equations by factorising the quadratic expression.

a) $b^2 - 25 = 0$ b) $y^2 - 144 = 0$ c) $64 - g^2 = 0$
d) $d^2 - 16 = 0$ e) $4x^2 = 100$ f) $36 = x^2$
g) $4n^2 = 25$ h) $2.25 - x^2 = 0$ i) $4x^2 - 48 = 16$

5 Solve these equations by factorising the quadratic expression.

a) $3x^2 - 10x + 8 = 0$ b) $2x^2 - 11x + 5 = 0$ c) $4g^2 + 16g + 15 = 0$
d) $6y^2 + 7y + 2 = 0$ e) $15 + 2b - b^2 = 0$ f) $3z^2 - 5z - 2 = 0$
g) $3m^2 - 12m - 15 = 0$ h) $6a^2 + 3a - 63 = 0$ i) $12y^2 - 25y + 12 = 0$

6 Solve these equations by rearranging and then factorising the quadratic expression.

a) $4y^2 = 7y$ b) $n^2 - 10n = 24$ c) $5y^2 = y(y + 3)$
d) $x^2 = 2x + 15$ e) $7a + a^2 = 18$ f) $7 = 8m - m^2$
g) $p(8 - p) = 16$ h) $x^2 - 5x + 6 = 3 - x$ i) $4 - 3b = b^2$
j) $15 - 7m = 2m^2$ k) $12a^2 = 35 + 16a$ l) $8x^2 = 3 - 2x$
m) $x(x - 4) = 12$ n) $x(3x - 1) = 2$ o) $(2x - 3)(3x - 4) = 15$

B Writing an equation from a solution set

If we have the solution set for a given quadratic equation, we can use this to write the factors, and then expand the factors to give the full original equation.

In general, if the solution set is $\{a, b\}$, then we know that the factors of the quadratic expression are $(x - a)$ and $(x - b)$.

So the quadratic equation will be $(x - a)(x - b) = 0$ or
$x^2 - ax - bx + ab = 0$

Examples

a) Find an equation that has a solution set of $\{2, 5\}$.

$$x = 2 \quad or \quad x = 5$$
$$x - 2 = 0 \quad or \quad x - 5 = 0$$
$$(x - 2)(x - 5) = 0$$
$$x^2 - 7x + 10 = 0$$

b) Find an equation that has a solution set of $\{\frac{3}{4}, -2\}$. Write your equation with whole number coefficients.

$$x = \frac{3}{4} \quad or \quad x = -2$$
$$4x = 3 \quad or \quad x + 2 = 0$$
$$4x - 3 = 0$$
$$(4x - 3)(x + 2) = 0$$
$$4x^2 + 5x - 6 = 0$$

Exercise 2

1 Find an equation (with whole number coefficients) for each solution set given below.

a) $\{4, 5\}$ b) $\{1, 2\}$ c) $\{9, -4\}$ d) $\{-2, 8\}$ e) $\{\frac{3}{4}, -2\}$

f) $\{-\frac{2}{3}, 1\}$ g) $\{\frac{1}{2}, -2\}$ h) $\{\frac{2}{5}, -1\}$ i) $\{0, 2\}$ j) $\{-3, 0\}$

C Solving problems using quadratic equations

You can solve many practical problems by using the maths you have learned. The information in the problem must be 'translated' into one or more mathematical equations. We can use algebra to solve these equations and then we can translate the maths solution back into words to find the answer to the problem.

We have already used this method in Coursebook 2 Unit 2 to solve problems with linear equations (where the highest power of the unknown variable is 1).

Some problems can only be solved by using quadratic equations, but the general approach is the same.

Here are some steps that will help you to solve most of the problems.

Step 1 Work in an organised manner. Read the whole problem first.
Step 2 Think carefully about the problem.
Write down what is given, and what you must find.
Step 3 Draw a diagram if appropriate; it helps to 'see' the problem.
Step 4 Choose a variable (e.g. x) to stand in place of what you must find, or the value connected with what you must find.
Step 5 Change the information in the problem from words into a maths sentence using symbols so that you can write an equation.
Step 6 Solve the equation to find the values of the unknown variable.
Remember: There are two solutions to a quadratic equation. One of the solutions may not make sense as a solution to the problem (e.g. negative, fractional, too large or too small numbers).
Step 7 Translate the result back into words to answer the question asked.
Step 8 Check that the answer is sensible (e.g. distances cannot be negative).

Example

The difference of two numbers is 3. Their product is 10.
What are the two numbers?

Let the smaller number be x.
So the larger number will be $x + 3$ (because their difference is 3).
Their product is 10, so

$$x(x + 3) = 10$$
$$x^2 + 3x = 10$$
$$x^2 + 3x - 10 = 0$$
$$(x + 5)(x - 2) = 0$$

So *either* $\quad x + 5 = 0 \quad or \quad x - 2 = 0$
$$x = -5 \quad or \quad x = 2$$

If the smaller number is 2, then the larger number will be $2 + 3 = 5$.
If the smaller number is -5, then the larger number will be $-5 + 3 = -2$.

So we have two solution sets:
The numbers can be $\quad$ 2 and 5 $\quad or \quad$ -5 and -2

Check: $\qquad 5 - 2 = 3 \quad$ and $\qquad 5 \times 2 = 10$
$\qquad -2 - (-5) = 3 \quad$ and $\quad (-2) \times (-5) = 10$

Example

Three numbers are in this sequence:
x, $x + 3$, $x + 6$.
The sum of the squares of the two smaller numbers is equal to the square of the larger number.
What are the three numbers?

We are told that:

$$(x)^2 + (x + 3)^2 = (x + 6)^2$$
$$x^2 + x^2 + 6x + 9 = x^2 + 12x + 36$$
$$x^2 + x^2 - x^2 + 6x - 12x + 9 - 36 = 0$$
$$x^2 - 6x - 27 = 0$$
$$(x - 9)(x + 3) = 0$$

So *either* $x - 9 = 0$ *or* $x + 3 = 0$
$$x = 9 \quad or \quad x = -3$$

If $x = 9$ then the three numbers will be:
9, $9 + 3 = 12$ and $9 + 6 = 15$
If $x = -3$ then the three numbers will be:
-3, $-3 + 3 = 0$, and $-3 + 6 = 3$

So we will have two solution sets:
The numbers can be   9, 12, 15
                or $-3, 0, 3$

Check: $9^2 + 12^2 = 81 + 144 = 225 = 15^2$
$(-3)^2 + 0^2 = 9 = 3^2$

Example

The sides of a square are $(x + 3)$ cm long. The area of this square equals the area of a rectangle with a length of $(2x + 3)$ cm and a width of $(3x - 5)$ cm.
Write an equation to find values for x and use these to find the area of the square.

Area of rectangle = Area of square
$$(2x + 3)(3x - 5) = (x + 3)(x + 3)$$
$$6x^2 - x - 15 = x^2 + 6x + 9$$
$$5x^2 - 7x - 24 = 0$$
$$(5x + 8)(x - 3) = 0$$
So *either* $x - 3 = 0$ *or* $5x + 8 = 0$
$$x = 3 \quad or \quad x = -\frac{8}{5}$$

||||➡

When $x = -\frac{8}{5}$, the width of the rectangle is

$$3x - 5 = 3\left(-\frac{8}{5}\right) - 5 = -4\frac{4}{5} - 5 = -9\frac{4}{5}$$

The width of a rectangle cannot be negative, so the value of $x = -\frac{8}{5}$ does not make sense.

So the only solution to this problem is $x = 3$.
This means that the side of the square = $x + 3 = 6$
The area of the square is $36\,\text{cm}^2$.

Check: Area of rectangle
$$= ([2 \times 3] + 3)([3 \times 3] - 5)$$
$$= 9 \times 4 = 36$$

which is the same as the area of the square.

Example   In a garden, a rectangle of grass is $10\,\text{m}$ long and $8\,\text{m}$ wide. The grass is surrounded by a path that is $x\,\text{m}$ wide. The total area of the grass and the path together is $143\,\text{m}^2$.
Calculate the width of the path (x).

The length of the grass plus the path on both sides is
$x + 10 + x = (2x + 10)$
The width of the grass plus the path on both sides is
$x + 8 + x = (2x + 8)$

Draw a diagram.

 length $\times$ width $=$ total area
$(2x + 10)(2x + 8) = 143$
$\quad 4x^2 + 36x + 80 = 143$
$\quad 4x^2 + 36x - 63 = 0$
$(2x - 3)(2x + 21) = 0$
So *either* $2x - 3 = 0$ *or* $2x + 21 = 0$
$$x = 1\frac{1}{2} \quad or \quad\quad\quad x = -10\frac{1}{2}$$

The width of the path cannot be negative, so the value of $x = -10\frac{1}{2}$ does not make sense.
The only solution to this problem is $x = 1\frac{1}{2}$,
or, in words, the width of the path is $1.5\,\text{m}$.

Check: Total area of grass and the path
$$= (10 + [2 \times 1.5]) \times (8 + [2 \times 1.5])$$
$$= (10 + 3) \times (8 + 3) = 13 \times 11 = 143$$

Example

A cross-country running course is in the shape of a right-angled triangle. The length of the course is 3 km. The longest leg of the course (the hypotenuse of the triangle) is 1.25 km long. What are the lengths of the other two shorter legs?

Draw a diagram.

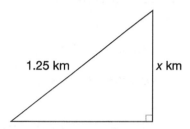

1.25 km x km

Let the length of one of the shorter legs be x km.
Then the length of the other shorter leg will be:
$3 - (1.25 + x) = 1.75 - x$

Now we can use Pythagoras' theorem.
$$x^2 + (1.75 - x)^2 = (1.25)^2$$
$$x^2 + 3.0625 - 3.5x + x^2 = 1.5625$$
$$2x^2 - 3.5x + 1.5 = 0$$
Multiply both sides by 10 and factorise the quadratic expression.
$$20x^2 - 35x + 15 = 0$$
$$5(4x^2 - 7x + 3) = 0$$
$$5(4x - 3)(x - 1) = 0$$
So *either* $4x - 3 = 0$, which gives $x = \frac{3}{4}$ or 0.75,
and the length of the other short leg is $(1.75 - 0.75) = 1$ km,
or $x - 1 = 0$, which gives $x = 1$,
and the length of the other short leg is $(1.75 - 1) = 0.75$ km.
These two solutions are both the same.
The lengths of the two shorter legs are 1 km and 0.75 km.

Check: $0.75 + 1 + 1.25 = 3$

and

$$(0.75)^2 + (1)^2$$
$$= 0.5625 + 1$$
$$= 1.5625$$
$$= (1.25)^2$$

Exercise 3

1 Eight more than the square of a number is the same as 6 times the number.
Find the number.

2 Seven less than 4 times the square of a number is 18.
Find the number.

3 Find two consecutive integers that have a product of 56.

4 The sum of the squares of two consecutive integers is 41.
Find the integers.

5 Find two consecutive odd integers so that the square of the first, added to 3 times the second, is 24.

6 Find two consecutive even integers so that the sum of their squares is 164.

7 Find two positive integers if their difference is 5 and the sum of their squares is 193.

8 The difference between two numbers is 9 and the product of the numbers is 162.
Find the two numbers.

9 The difference between two numbers is 3. The square of the smaller number is equal to 4 times the larger number.
Find the two numbers.

10 Two positive integers differ by 5 and the square of their sum is 169.
Find the numbers.

11 The length of a rectangle is 3 cm more than the width. The area is 70 cm^2.
Find the length and width of the rectangle.

12 The length of a photograph is 1 cm less than twice the width. The area is 45 cm^2.
Find the length and width of the photograph.

13 A square field is made bigger by adding 5 m to the length and 2 m to the width.
The area of the new (bigger) field is 130 m^2.
Find the length of a side of original field.

14 A rectangular garden measures 3 m by 10 m. The length and width are both increased by the same amount.
The area of the new garden is double the area of the old garden.
Find the length and width of the new garden.

15 A rug measures $4\,m$ by $6\,m$. It is placed in the centre of a room of area $48\,m^2$.
There is now a strip of bare wooden floor that is the same width all around the carpet.
How big is this room?

16 The length of a rectangle is $5\,cm$ longer than the width. The area is $66\,cm^2$.
Find the perimeter of the rectangle.

17 The length of a rectangle is $(4x + 7)\,cm$ and the width is $(5x - 4)\,cm$.
The area of the rectangle is $209\,cm^2$.
Find the perimeter of this rectangle.

18 The sides of rectangle A are $5x\,cm$ and $(4x + 2)\,cm$.
The sides of rectangle B are $(6x + 3)\,cm$ and $(3x + 1)\,cm$.
The area of A is equal to the area of B.
Find the perimeter of each rectangle.

19 The perpendicular height of a triangle is $(2x - 5)\,cm$ and the base measures $(x + 3)\,cm$. The area of this triangle is $20\,cm^2$.
Find the length of the base.

20 The perpendicular height of a trapezium is $(3x - 4)\,cm$.
The lengths of the two parallel sides are $(x + 9)\,cm$ and $(x + 3)\,cm$.
The area of the trapezium is $80\,cm^2$.
Calculate the lengths of the two parallel sides.

21 Michael is 5 years older than his brother, Peter. The product of their ages is 234.
How old is Peter?

D Completing a perfect square

We have learned that a perfect square in algebra is made by multiplying the **whole** of an expression by itself.
Choose a and b to be constant numbers, and x to be a variable:
Then, for the expression $(ax + b)$, its perfect square will be
$(ax + b)^2 = a^2x^2 + 2abx + b^2$

Now, compare the expression $(ax + b)$ with its square
$a^2x^2 + 2abx + b^2$.

- The first term of the perfect square expression is the same as the square of the first term of the original expression as $(ax)^2 = a^2x^2$.
- The first term of an expression that is a perfect square is also **always** a perfect square.

- The last term of an expression that is a perfect square is also **always** a perfect square. In fact it is the square of the constant term, b, in the original expression.
- The coefficient of the middle term of the perfect square is $2ab$. We can write this as:
  2 × (the coefficient of x in the original expression) × (the constant term of the original expression)

Completing the square when the coefficient of the x^2 term is 1

We will start by looking at the special case when $a = 1$ in our expression $(ax + b)$, because it is easier to see the pattern in this case. The perfect square will be $x^2 + 2bx + b^2$

Now, we can say that the constant term of the perfect square (b^2) is the same as:

$$\text{(half the coefficient of the } x \text{ term)}^2 = \left(\tfrac{1}{2} \times 2b\right)^2$$

Check that this rule is true by looking at these examples.

$(x + 1)^2 = x^2 + 2x + 1 \qquad (x - 1)^2 = x^2 - 2x + 1$
$(x + 2)^2 = x^2 + 4x + 4 \qquad (x - 2)^2 = x^2 - 4x + 4$
$(x + 3)^2 = x^2 + 6x + 9 \qquad (x - 3)^2 = x^2 - 6x + 9$

If we have an expression such as $x^2 + 8x$ where the coefficient of x^2 is 1 and we have the x term, then we can use this rule to find the constant term that will make the expression into a perfect square.

Examples

Find a constant term to add to each of these expressions so that it will become a perfect square. Then write your answer in the form $(x + b)^2$ where b is a number.

In each case, the constant term will be (half the coefficient of x)^2.

a) $x^2 + 4x$

Half the coefficient of x is $\tfrac{4}{2}$, so we add the square of this to the original expression.
$$x^2 + 4x + \left(\tfrac{4}{2}\right)^2 = \left(x + \tfrac{4}{2}\right)^2 = (x + 2)^2$$

b) $x^2 - 6x$

Half the coefficient of x is $\tfrac{-6}{2}$, so we add the square of this to the original expression.
$$x^2 - 6x + \left(\tfrac{-6}{2}\right)^2 = \left(x - \tfrac{6}{2}\right)^2 = (x - 3)^2$$

c) $x^2 + 5x$

Half the coefficient of x is $\frac{5}{2}$, so we add the square of this to the original expression.

$x^2 + 5x + \left(\frac{5}{2}\right)^2 = \left(x + \frac{5}{2}\right)^2 = (x + 2.5)^2$

d) $x^2 - \frac{1}{3}x$

Half the coefficient of x is $\left(\dfrac{-\frac{1}{3}}{2}\right)$, so we add the square of this to the original expression.

$x^2 - \frac{1}{3}x + \left(\dfrac{-\frac{1}{3}}{2}\right)^2 = \left(x - \dfrac{\frac{1}{3}}{2}\right)^2 = \left(x - \frac{1}{6}\right)^2$

This process of adding a constant term to a quadratic expression to make it into a perfect square is called **completing the square**.

If the x^2 term of any quadratic expression is a perfect square, then we can write this expression as a **perfect square term** and a **constant term**. This is sometimes called the **completed square form**.
In general it is written as $(x + b)^2 + c$ (where b and c are numbers).
For example, the quadratic expression $x^2 + 8x + 18$ is **not** a perfect square but the x^2 term is a perfect square. We have learned how to change the first part $(x^2 + 8x)$ into a perfect square in the form $(x + b)^2$:
The coefficient of x is 8. Half of 8 is 4. So the perfect square from this part will be $(x + 4)^2$.
That will give us $\qquad (x + 4)^2 = x^2 + 8x + 16$
But our original expression is $x^2 + 8x + 18$, which we can write as
$x^2 + 8x + 16 + 2 = (x + 4)^2 + 2$

So $\quad x^2 + 8x + 18 = (x + 4)^2 + 2 \qquad\qquad$ (completed square form)

Examples

Write these quadratic expressions in the completed square form $(x + b)^2 + c$.

a) $x^2 + 12x + 14$

The coefficient of x is 12.
Half of 12 is 6.
$\qquad\qquad (x + 6)^2 = x^2 + 12x + 36$
$x^2 + 12x + 14 = x^2 + 12x + 36 - 22$
$\qquad\qquad\quad = (x + 6)^2 - 22$

b) $x^2 - 6x - 7$

The coefficient of x is -6.
Half of -6 is -3.
$$(x - 3)^2 = x^2 - 6x + 9$$
$$x^2 - 6x - 7 = x^2 - 6x + 9 - 16$$
$$= (x - 3)^2 - 16$$

Completing the square when the coefficient of the x^2 term is bigger than 1

We know that the x^2 term must be a perfect square if we want to change the whole expression into a perfect square. This means that the only coefficients of x^2 that are bigger than 1 and will work with our rule are the square numbers: 4, 9, 16, 25, … and so on.

Look at the following perfect squares.

$(2x + 1)^2 = 4x^2 + 4x + 1$ $(2x - 1)^2 = 4x^2 - 4x + 1$
$(3x + 2)^2 = 9x^2 + 12x + 4$ $(3x - 2)^2 = 9x^2 - 12x + 4$
$(4x + 3)^2 = 16x^2 + 24x + 9$ $(4x - 3)^2 = 16x^2 - 24x + 9$

Can you see the pattern? In general, we can say:
$(ax + b)^2 = a^2x^2 + 2abx + b^2$ (where a and b are numbers).

We can use this to help us to change any quadratic expression (where the coefficient of x^2 is a perfect square number) into the completed square form.

1   We can find the value of a because we know
    $a = \sqrt{\text{coefficient of the } x^2 \text{ term}}$.

2   We can also find the value of b because we know that the coefficient of the x term $= 2ab$.

Examples

Find a constant term to add to each of these expressions so that it will become a perfect square. Then write your answer in the form $(ax + b)^2$ where a and b are numbers.

a) $16x^2 + 24x$

We know that $a = \sqrt{16} = 4$
We also know that $2ab = 24$
So $2 \times 4 \times b = 24$
and $b = 3$
The constant term to make a perfect square will be $b^2 = 3^2 = 9$.
The perfect square is:
$16x^2 + 24x + 9 = (4x + 3)^2$

b) $9x^2 - 30x$

We know that $a = \sqrt{9} = 3$
We also know that $2ab = 30$
So $\quad 2 \times 3 \times b = -30$
and $\qquad\qquad b = -5$
The constant term to make a perfect square will be
$b^2 = (-5)^2 = 25$.
The perfect square is:
$9x^2 - 30x + 25 = (3x - 5)^2$

As before, we can write any expression that has a perfect square coefficient of the x^2 term in the completed square form: $(ax + b)^2 + c$ (where a, b and c are numbers).

As an example, we will write the quadratic expression $4x^2 + 28x + 51$ in the completed square form, $(ax + b)^2 + c$.

So we must find the values of a, b and c.

We already know how to work out the values of a and b:

$a = \sqrt{4} = 2$
$2ab = 28$, so $2 \times 2 \times b = 28$ and $b = 7$

This gives us the perfect square $(2x + 7)^2$.
But $(2x + 7)^2 = 4x^2 + 28x + 49$
So we can write our starting expression as

$4x^2 + 28x + 51 = 4x^2 + 28x + 49 + 2$

In the completed square form we have

$4x^2 + 28x + 51 = (2x + 7)^2 + 2$

Example

Change $25x^2 + 30x + 7$ into the completed square form $(ax + b)^2 + c$.

$a = \sqrt{25} = 5$
$2ab = 30$, so $2 \times 5 \times b = 30$ and $b = 3$
This gives us the perfect square $(5x + 3)^2$.
But $(5x + 3)^2 = 25x^2 + 30x + 9$
So we can write our starting expression as

$25x^2 + 30x + 7 = 25x^2 + 30x + 9 - 2$

In the completed square form we have

$25x^2 + 30x + 7 = (5x + 3)^2 - 2$

Exercise 4

1 Add a constant term to each expression to make it into a perfect square.

a) $x^2 + 6x$ b) $x^2 + 10x$ c) $x^2 - 8x$ d) $x^2 - 9x$

e) $x^2 + 7x$ f) $x^2 + \frac{2}{5}x$ g) $x^2 - \frac{4}{3}x$ h) $x^2 + 2.4x$

2 Change each of these quadratic expressions into the completed square form.

a) $x^2 + 12x - 5$ b) $x^2 + 22x + 97$

c) $x^2 - 14x + 7$ d) $x^2 - 2x + 19$

e) $x^2 + 10x + 29.25$ f) $x^2 + x + 3$

g) $x^2 - 7x + 10$ h) $x^2 + 1.8x - 1.19$

3 Add a constant term to each expression to make it into a perfect square.

a) $36x^2 + 12x$ b) $4x^2 + 36x$ c) $25x^2 - 50x$ d) $64x^2 - 48x$

e) $9x^2 + 3x$ f) $16x^2 + 2x$ g) $49x^2 - 7x$ h) $25x^2 + 13x$

4 Change each of these quadratic expressions into the completed square form.

a) $9x^2 + 12x - 12$ b) $4x^2 + 24x + 11$

c) $16x^2 - 56x + 33$ d) $25x^2 - 90x - 19$

E Solving quadratic equations by completing the square

We have learned how to change some expressions into the completed square form. Now we will use this method to help us solve quadratic equations.

So far, we have used the zero product theorem to solve equations where the quadratic expression can be factorised.

If the quadratic expression cannot be factorised, but the x^2 term is a perfect square, we can use the method of completing the square to solve the equation.

We first change the quadratic expression into the completed square form. Then we move the constant term to the right-hand side, leaving the perfect square on the left, so the equation will look something like this:

$(ax + b)^2 = \text{constant term}$

We then take the **square root** of each side to find the values of x.

The value of the constant term on the right-hand side of the equation will tell us what sort of numbers our values of x will be:

- If the constant term is a **positive perfect square**, then the values of x will be rational numbers.
- If the constant term is any other **positive rational number**, then the values of x will be real numbers.
- If the constant term is any **negative real number**, then the values of x will be imaginary and we will have no real values for x that are true for the equation.

For example, look at the equation $x^2 + 4x - 5 = 0$
This is not a perfect square, but the x^2 term is a perfect square, so we can write the expression in the completed square form:
$(x + 2)^2 - 4 - 5 = 0$ or $(x + 2)^2 - 9 = 0$
Now move the constant term to the right-hand side:
$(x + 2)^2 = 9$
Then take the square root of both sides:
$(x + 2) = \pm\sqrt{9} = \pm 3$ (a square root gives both a positive and negative value)

So the values of x will be: $x + 2 = 3$ or $x + 2 = -3$
$$x = 1 \quad or \quad x = -5$$
The solution set is $\{1, -5\}$.

We could also solve this equation using factorisation and the zero product theorem:

$$x^2 + 4x - 5 = 0$$
$$(x + 5)(x - 1) = 0$$
So $x + 5 = 0$ or $x - 1 = 0$
and $x = -5$ or $x = 1$ (Solution set: $\{-5, 1\}$)

It is important to note that we find the same values of x by using both methods.

Examples   Solve these quadratic equations by completing the square.

a) $x^2 - 6x = 66$

This quadratic expression is not a perfect square, but the coefficient of x^2 is a perfect square (1).
The first step is to change the expression on the LHS into the completed square form.
$$x^2 - 6x = 66$$
$$(x - 3)^2 - 9 = 66$$

Now take the constant terms to the RHS.
$$(x - 3)^2 = 75$$

Then take the square root of both sides. $\sqrt{(x - 3)^2} = \pm\sqrt{75}$

(Remember: a square root can be positive or negative)
$$(x - 3) = \pm 8.66 \text{ (2 d.p.)}$$
So *either* $x - 3 = 8.66$ *or* $x - 3 = -8.66$
$$x = 11.66 \quad or \qquad x = -5.66$$

Solution set: $\{11.66, -5.66\}$

b) $4x^2 + 16x - 9 = 0$

This quadratic expression is not a perfect square, but the coefficient of x^2 is a perfect square (4).
The first step is to change the expression on the LHS into the completed square form.
$$4x^2 + 16x - 9 = 0$$
$$(2x + 4)^2 - 16 - 9 = 0$$

Now take the constant terms to the RHS: $(2x + 4)^2 = 25$
Then take the square root of both sides.

$$\sqrt{(2x + 4)^2} = \pm\sqrt{25}$$
$$(2x + 4) = \pm 5$$

So *either* $2x + 4 = 5$ or $2x + 4 = -5$
$$x = \tfrac{1}{2} \quad or \qquad x = -4\tfrac{1}{2}$$

Solution set: $\left\{\tfrac{1}{2}, -4\tfrac{1}{2}\right\}$

If the coefficient of the x^2 term is **not** a perfect square, we can still use this method of completing the square to solve the equation. We must first divide the whole equation by the coefficient of the x^2 term so that it will become 1, and we can then continue as before:

Examples   Solve these quadratic equations by completing the square.

a) $2x^2 - 12x - 9 = 0$

This quadratic expression is not a perfect square, and neither is the coefficient of x^2.
First divide both sides of the equation by the same number so that the coefficient of x^2 becomes 1.
(Divide by 2) $x^2 - 6x - 4.5 = 0$

Now we can complete the square on the LHS.
$$x^2 - 6x - 4.5 = 0$$
$$[x^2 - 6x + 9 - 9] - 4.5 = 0$$
$$[(x - 3)^2 - 9] - 4.5 = 0$$
$$(x - 3)^2 - 13.5 = 0$$

Now take the constant terms to the RHS: $(x - 3)^2 = 13.5$
Then take the square root of both sides.

$$\sqrt{(x - 3)^2} = \pm\sqrt{13.5}$$
$$(x - 3) = \pm 3.67 \text{ (3 s.f.)}$$

So *either* $x - 3 = 3.67$ *or* $x - 3 = -3.67$
$$x = 6.67 \quad or \quad x = -0.67$$

(The answers are correct to 3 significant figures.)

Solution set: $\{6.67, -0.67\}$

b) $3x^2 - 12x + 7 = 0$

This quadratic expression is not a perfect square, and neither is the coefficient of x^2.
First divide both sides of the equation by the same number so that the coefficient of x^2 becomes 1:

(Divide by 3) $x^2 - 4x + 1\frac{2}{3} = 0$

Now we can complete the square on the LHS.
$$x^2 - 4x + 1\frac{2}{3} = 0$$
$$[x^2 - 4x + 4 - 4] + 1\frac{2}{3} = 0$$
$$[(x - 2)^2 - 4] + 1\frac{2}{3} = 0$$
$$(x - 2)^2 - 1\frac{2}{3} = 0$$

Now take the constant terms to the RHS: $(x - 2)^2 = 1\frac{2}{3}$

Then take the square root of both sides.
$$\sqrt{(x - 2)^2} = \pm\sqrt{1\frac{2}{3}}$$
$$(x - 2) = \pm 1.29 \text{ (2 d.p.)}$$
So *either* $x - 2 = 1.29$ *or* $x - 2 = -1.29$
$$x = 3.29 \quad or \quad x = 0.71$$
(The answers are correct to 2 decimal places.)

Solution set: $\{3.29, 0.71\}$

Solving quadratic equations by completing the square

1 Write the equation with all terms on the LHS.
2 Check the coefficient of the x^2 term.
   a) If it is 1, continue to step **3**.
   b) If it is another perfect square number but not equal to 1, also continue to step **3**.
   c) If it is any number that is not a perfect square, then divide the **whole** equation by this number so that the coefficient of the x^2 term now becomes 1.
3 Change the quadratic expression on the LHS into the completed square form.
4 Take any constant number terms to the RHS of the equation and combine them.
   (The equation should now be in the form:
   perfect square term = constant number term.)
5 Take the **square root** of both sides of the equation.
   (Remember: taking the square root of a number gives a positive and a negative answer.)
6 This gives two equations with two possible values of x as the solution set of the equation.
7 Check both solutions to see if they make sense if you are working out a word problem.
   (For example, you cannot have negative length.)

 Exercise 5

 1 Solve each of these equations by completing the square. If the answers are not whole numbers, give them correct to 2 decimal places.

a) $x^2 + 14x + 48 = 0$ b) $x^2 + 20x + 96 = 0$ c) $x^2 + 63 = 16x$
d) $x^2 - 2x = 195$ e) $9x^2 - 12x = 60$ f) $16x^2 + 56x + 13 = 0$
g) $x^2 + 12 = 9x$ h) $x^2 = 3x + 10$ i) $x^2 + 9x + 11 = 0$
j) $x^2 + 11x + 15 = 0$ k) $3x^2 - 18x - 21 = 0$ l) $5x^2 = 4x - 2$
m) $2x^2 + 5x = 3$ n) $7x^2 - 28x + 15 = 0$ o) $2x^2 + 3x - 4 = 0$
p) $3x^2 + x = 2$ q) $3x^2 + 5x - 2 = 0$ r) $x^2 - 6x - 16 = 0$
s) $x^2 = 16x + 10$ t) $5x^2 - 8x - 30 = 0$

The quadratic formula

When a quadratic expression does not factorise, it is often possible to use the method of completing the square to solve the quadratic equation.

However, sometimes this is a very long process or is not possible at all using this method. For example, how would you take the square root of both sides if your completed square was $(2x - 3)^2 = -9$?

The method of completing the square can be followed for the general form of any quadratic equation: $ax^2 + bx + c = 0$ where a, b and c are any rational numbers.
This gives a **formula** that we can use for **any quadratic equation** (whether it factorises or not).

Taking this equation, we follow the steps for completing the square (in blue here):

1 Write the equation with all terms on the LHS:
$$ax^2 + bx + c = 0$$
2 Check the coefficient of the x^2 term.
  a) If it is 1, continue to step **3**.
  b) If it is another perfect square number but not equal to 1, also continue to step **3**.
  c) If it is any number that is not a perfect square, then divide the whole equation by this number so that the coefficient of the x^2 term now becomes 1.
  As we don't know exactly what number 'a' is, we must follow part c) to give:
$$x^2 + \frac{b}{a}x + \frac{c}{a} = 0$$
3 Change the quadratic expression on the LHS into the completed square form:

$$\left[x^2 + \frac{b}{a}x + \left(\frac{\frac{b}{a}}{2}\right)^2 \right] - \left(\frac{\frac{b}{a}}{2}\right)^2 + \frac{c}{a} = 0$$

$$\left[\left(x + \frac{\frac{b}{a}}{2}\right)^2 \right] - \left(\frac{\frac{b}{a}}{2}\right)^2 + \frac{c}{a} = 0$$

$$\left[\left(x + \frac{b}{2a}\right)^2 \right] - \left(\frac{b}{2a}\right)^2 + \frac{c}{a} = 0$$

4 Take any constant number terms to the RHS of the equation and combine them:

$$\left(x + \frac{b}{2a}\right)^2 = \left(\frac{b}{2a}\right)^2 - \frac{c}{a}$$

Now we can simplify the RHS of the equation:

$$\left(x + \frac{b}{2a}\right)^2 = \frac{b^2}{4a^2} - \frac{c}{a}$$

$$\left(x + \frac{b}{2a}\right)^2 = \frac{b^2 - 4ac}{4a^2}$$

5 Take the square root of both sides of the equation:

$$x + \frac{b}{2a} = \pm\sqrt{\frac{b^2 - 4ac}{4a^2}}$$

Now we can simplify the denominator on the RHS because $4a^2$ is a perfect square:

$$x + \frac{b}{2a} = \pm\frac{\sqrt{b^2 - 4ac}}{2a}$$

Finally we can solve for x:

$$x = -\frac{b}{2a} \pm \frac{\sqrt{b^2 - 4ac}}{2a}$$

Simplify the two fractions on the RHS over the same denominator:

$$x = \frac{-b \pm \sqrt{b^2 - 4ac}}{2a}$$

This formula gives us the solution for **any** quadratic equation. You must learn it.

> The solution for any quadratic equation $ax^2 + bx + c = 0$ is
>
> $$x = \frac{-b \pm \sqrt{b^2 - 4ac}}{2a}$$
>
> where a, b and c are numbers.

G Using the quadratic formula to solve quadratic equations

Examples

Solve the following quadratic equations using the formula. Give your answers correct to 2 d.p.

a) $x^2 + 4x + 2 = 0$

Compare $x^2 + 4x + 2 = 0$ with $ax^2 + bx + c = 0$
So $a = 1$, $b = +4$ and $c = +2$

Using the formula: $x = \dfrac{-b \pm \sqrt{b^2 - 4ac}}{2a}$

$= \dfrac{-4 \pm \sqrt{4^2 - 4(1)(2)}}{2(1)}$

$= \dfrac{-4 \pm \sqrt{16 - 8}}{2} = \dfrac{-4 \pm \sqrt{8}}{2}$

$= \dfrac{-4 + 2.828}{2}$ or $\dfrac{-4 - 2.828}{2}$

$= \dfrac{-1.172}{2}$ or $\dfrac{-6.828}{2}$

$= -0.586$ or -3.414

$x = -0.59$ or -3.41 (to 2 d.p.)

b) $6x = 3 - 5x^2$

Write all terms on the RHS: $5x^2 + 6x - 3 = 0$
Compare this with $ax^2 + bx + c = 0$
So $a = 5$, $b = +6$ and $c = -3$

Using the formula: $x = \dfrac{-b \pm \sqrt{b^2 - 4ac}}{2a}$

$= \dfrac{-6 \pm \sqrt{6^2 - 4(5)(-3)}}{2(5)}$

$= \dfrac{-6 \pm \sqrt{36 + 60}}{10} = \dfrac{-6 \pm \sqrt{96}}{10}$

$= \dfrac{-6 + 9.798}{10}$ or $\dfrac{-6 - 9.798}{10}$

$= \dfrac{3.798}{10}$ or $\dfrac{-15.798}{10}$

$= 0.380$ or -1.580

$x = 0.38$ or -1.58 (to 2 d.p.)

Exercise 6

 1 Solve these quadratic equations using the quadratic formula.
Give your answers correct to 2 decimal places.

a) $x^2 + 7x + 5 = 0$ b) $x^2 + 3 = 8x$ c) $x^2 - 3x = 2$

d) $x^2 + 9x + 4 = 0$ e) $x^2 + 5x + 3 = 0$ f) $x^2 = 10x + 10$

g) $x^2 = 3x + 5$ h) $3x^2 + 1 = 5x$ i) $2x^2 = 4x + 5$

j) $3x^2 + 23x + 8 = 0$ k) $3x^2 + 6x = 2$ l) $2x^2 + 6x + 3 = 0$

m) $6x^2 = 17$ n) $2x^2 = 3x + 6$ o) $7x^2 = x + 12$

p) $2 = x + 5x^2$

We have learned **three ways** to solve a quadratic equation:

1 Factorise the quadratic expression and use the zero product theorem.

2 Complete the square.

3 Use the quadratic formula.

Now look at some more word problems, choosing one of these three methods to solve the equation.

Remember that you can often use more than one of the methods for any quadratic equation, but each method should give the same answers for any equation.

Exercise 7

 In each question, choose a suitable method to solve the quadratic equation.
Give the answers correct to 1 decimal place if necessary.

1 The length of a rectangle is 4 m more than the width. The area of the rectangle is 45 m².
Calculate the length and width of this rectangle.

2 A garden is in the shape of a rectangle. The garden is 4 m longer than it is wide.
The area of the garden is 71 m².
Calculate the length and width of the garden.

3 A flowerbed is in the shape of a right-angled triangle. The longest side of the flowerbed measures 13 m. One of the shorter sides is 7 m longer than the third one.
Calculate the perimeter of the flowerbed.

4 The length of a photograph is 1 cm less than twice the width.
The area is 28 cm².
Calculate the length and width of the photograph.

5 A garden pond is in the shape of a rectangle that measures 5 m by 3 m.
A stone path is built all around the pond. This path is the same width all the
way around.
The area of the pond and the path together is 39 m².
How wide is the path?

6 The width of a rectangular painting is 2 m less than the length.
The area of the painting is 20 m².
Calculate the length and width of the painting.

7 The perpendicular height of a triangle is 2 cm shorter than its base.
The area of the triangle is 15 cm².
Find the length of the base of this triangle.

8 A movie screen is 2 m longer than it is wide. The diagonal of the screen
measures 6 m.
Calculate the length and width of the screen.

Unit 4 Algebraic fractions

Key vocabulary

algebraic fraction	monomial	polynomial
denominator	numerator	

Revision

In Unit 3 of Coursebook 1 we studied fractions:

- We know that a fraction is made by **dividing** one number by another number.
- We write fractions as one number over another number (not with the division symbol, ÷).
- The number that is being divided is written at the top and is called the numerator.
- The number we are dividing by is written at the bottom and is called the denominator.

$$\text{Fraction} = \frac{\text{numerator}}{\text{denominator}} = \text{numerator} \div \text{denominator}$$

We also know that an **algebraic expression** contains **variables** (for unknown numbers).

An algebraic fraction has the same form as any other fraction, with numerator over denominator, but at least one of them is an algebraic expression containing a variable.

Here are some examples:

$$\frac{1}{x} \quad \frac{y}{3} \quad \frac{4}{a+7} \quad \frac{m^2-1}{9} \quad \frac{b+5}{b-5} \quad \frac{x^2-6x+9}{x^2-9}$$

All the rules we have already learned to simplify, add, subtract, multiply and divide fractions can also be used with algebraic fractions.

NOTE: Because a fraction is another way of writing **division**, we have already studied some algebraic fractions when we learned about division of algebraic expressions in Unit 9 of Coursebook 1 and Unit 8 of Coursebook 2, which you might like to look at again.

A # Simplifying algebraic fractions

We have already learned that all fractions can be simplified if the numerator and the denominator have a **common factor**.

To write a fraction in its **simplest form**, we divide both the numerator and the denominator by their highest common factor (HCF). This is called 'cancelling'.

So far, we have learned how to simplify algebraic fractions where the **denominator is a** monomial (an expression with only **one term**).

Simplify these algebraic fractions.

a) $\dfrac{(4a)^2b^3}{(-2ab^2)^3}$

First simplify the numerator and denominator as much as possible. This could mean that we need to expand brackets with an index outside, or to factorise a polynomial expression.

$$\frac{(4a)^2b^3}{(-2ab^2)^3} = \frac{16a^2b^3}{-8a^3b^6}$$

Now it is much easier to find the HCF of the numerator and the denominator.

$$\text{HCF} = 8a^2b^3$$

Divide both numerator and denominator by the HCF to give the simplest form (cancelling: multiply by $\frac{\text{HCF}}{\text{HCF}}$):

$$\frac{16a^2b^3}{-8a^3b^6} \times \frac{8a^2b^3}{8a^2b^3}$$

$$= \frac{2}{-ab^3} = \frac{-2}{ab^3} \quad or \quad -\frac{2}{ab^3}$$

b) $\dfrac{2a^3b - 8a^2b^2 - 10ab^3}{-2ab}$

$$\frac{2a^3b - 8a^2b^2 - 10ab^3}{-2ab} = \frac{2ab(a^2 - 4ab - 5b^2)}{-2ab}$$

$$= \frac{2ab(a + b)(a - 5b)}{-2ab}$$

HCF = $2ab$, so divide numerator and denominator by $2ab$.

$$= -(a + b)(a - 5b)$$

We can use the same steps to simplify any algebraic fraction:

- Simplify and/or factorise the expression in the numerator and the denominator.
- Find the HCF of the numerator and the denominator.
- Divide the numerator and denominator by the HCF (cancelling).
- Write the answer in its simplest form.

NOTE: Revise Unit 9 in Coursebook 2 on factorising algebraic expressions, and also Unit 7 in Coursebook 2 on indices.

Now we will use these simple steps to factorise algebraic fractions that have a polynomial in the denominator.

Examples

Simplify these algebraic fractions.

a) $\dfrac{xy + 3y}{2x + 6}$

Simplify and/or factorise the expression in the numerator and the denominator.

$$\frac{xy + 3y}{2x + 6} = \frac{y(x + 3)}{2(x + 3)}$$

Find the HCF of the numerator and the denominator.

$HCF = (x + 3)$

Divide the numerator and denominator by the HCF (cancelling).

$$\frac{y(x + 3)}{2(x + 3)} \times \frac{(x + 3)}{(x + 3)} = \frac{y}{2} = \tfrac{1}{2}y$$

b) $\dfrac{4x^3 + 12x^2y}{8x^4 + 24x^3y}$

Simplify and/or factorise the expression in the numerator and denominator.

$$\frac{4x^3 + 12x^2y}{8x^4 + 24x^3y} = \frac{4x^2(x + 3y)}{8x^3(x + 3y)}$$

Find the HCF of the numerator and the denominator.

$HCF = 4x^2(x + 3y)$

Divide the numerator and denominator by the HCF (cancel).

$$\frac{4x^2(x + 3y)}{8x^3(x + 3y)} \times \frac{4x^2(x + 3y)}{4x^2(x + 3y)} = \frac{1}{2x} \qquad \text{(This is \textbf{not} the same as } \tfrac{1}{2}x.)$$

c) $\dfrac{b^2 - a^2}{2a^2 + ab - 3b^2}$

Simplify and/or factorise the expression in the numerator and the denominator.

$$\dfrac{b^2 - a^2}{2a^2 + ab - 3b^2} = \dfrac{(b - a)(b + a)}{(2a + 3)(a - b)}$$
$$= \dfrac{-(a - b)(b + a)}{(2a + 3)(a - b)}$$

Find the HCF of the numerator and the denominator.

HCF $= (a - b)$

Divide the numerator and denominator by the HCF (cancelling).

$$\dfrac{-(a - b)(a + b)}{(2a + 3)(a - b)} \times \dfrac{(a - b)}{(a - b)}$$
$$= \dfrac{-(a + b)}{(2a + 3)}$$

d) $\dfrac{a^2 + 11ab + 18b^2}{a^2b + 9ab^2}$

Simplify and/or factorise the expression in the numerator and denominator.

$$\dfrac{a^2 + 11ab + 18b^2}{a^2b + 9ab^2} = \dfrac{(a + 2b)(a + 9b)}{ab(a + 9b)}$$

Find the HCF of the numerator and the denominator.

HCF $= (a + 9b)$

Divide the numerator and denominator by the HCF (cancelling).

$$\dfrac{(a + 2b)(a + 9b)}{ab(a + 9b)} \times \dfrac{(a + 9b)}{(a + 9b)}$$
$$= \dfrac{(a + 2b)}{ab}$$

Exercise 1

1 Simplify each of these algebraic fractions.

a) $\dfrac{2x + 10}{x + 5}$ b) $\dfrac{x + 3}{9 - x^2}$ c) $\dfrac{x^2 - 4}{x + 2}$

d) $\dfrac{5x - 20}{4 - x}$ e) $\dfrac{x^2 + 4x}{x^2 - 9x}$ f) $\dfrac{n^2 - 7n + 12}{n^2 - 2n - 3}$

g) $\dfrac{x^2 - 2x - 15}{10 - 2x}$ h) $\dfrac{4n + 28}{n^2 + 6n - 7}$ i) $\dfrac{-x^2 - 3x + 10}{25 - x^2}$

j) $\dfrac{2b^2 - 6b}{5b^2 - 15b}$ k) $\dfrac{-c^2 + 6c - 9}{c^2 + 5c - 24}$ l) $\dfrac{3b^2 + 15b}{2b^3 - 50b}$

m) $\dfrac{c^2 - 9d^2}{2c + 6d}$ n) $\dfrac{6b^3 - 24b^2}{b^2 + b - 20}$ o) $\dfrac{-c^2 + 2cd + 3d^2}{5c - 15d}$

p) $\dfrac{2x^3 - 7x^2 - 15x}{2x^2 - 5x - 12}$

2 Simplify each of these algebraic fractions.

a) $\dfrac{2x^2 - 18}{4x + 12}$ b) $\dfrac{3x^2 - 24x + 36}{2x^2 - x - 6}$ c) $\dfrac{-x^2 + 8x - 16}{x^3 - 4x^2}$

d) $\dfrac{2a^2b^2 + 4ab^2}{a^4b + 4a^3b}$ e) $\dfrac{8a^2b - 8b^3}{6a^2b + 12ab^2 + 6b^3}$ f) $\dfrac{10a^3b + 10a^2b}{4a^2b^3 + 2ab^3}$

g) $\dfrac{(b - 5)^3}{15 + 7b - 2b^2}$ h) $\dfrac{4a^3b^4(a^2 + a - 42)}{28a^4b^4(6 - a)}$ i) $\dfrac{a^5b^2(a^2 + 7a + 10)}{a^2b^4(a + 5)}$

j) $\dfrac{12a^5b^3(3 - b)}{4a^2b(b^2 + b - 12)}$

B # Multiplication and division of algebraic fractions

When we multiply and divide fractions, we find the final answer by dividing the numerator and denominator by their common factors (cancelling): for example, $\dfrac{20}{6} \times \dfrac{3}{4} = \dfrac{5}{2}$.

When we multiply and divide algebraic fractions, we use exactly the same method. Remember to **change** any ÷ sign into a × sign and **invert** the fraction that was directly after the ÷ sign.

It is not easy to see all the factors of the algebraic expressions in the numerator and the denominator. First we must factorise the algebraic expressions so that all their factors are clear and we can see which common factors we can cancel.

NOTE: When both the numerator and the denominator of all the algebraic fractions are **monomials**, the multiplication and division is exactly the same as we learned in Unit 7 of Coursebook 2 when we studied indices,

$$\left(\text{e.g. } \frac{3m^2p^4}{14m^3p^3} \times \frac{7m^4p^2}{12mp^3} = \frac{m^{2+4-3-1}p^{4+2-3-3}}{2 \times 4} = \frac{m^2}{8} \right)$$

Examples

Simplify these algebraic expressions and give the answers in their lowest terms.

a) $\dfrac{5 + 5y}{y - 3} \times \dfrac{6 - 2y}{8y + 8}$

First factorise all the algebraic expressions.

$$\frac{5 + 5y}{y - 3} \times \frac{6 - 2y}{8y + 8} = \frac{5(1 + y)}{y - 3} \times \frac{-2(y - 3)}{8(y + 1)}$$

Now cancel common factors [remember: $(y + 1) = (1 + y)$].

$$= \frac{5 \times -1}{4}$$

$$= -1\frac{1}{4}$$

b) $\dfrac{x^2 - 2xy}{x + y} \div \dfrac{x^2 - 4y^2}{x^2 + 2xy + y^2}$

First factorise all the algebraic expressions.

$$\frac{x^2 - 2xy}{x + y} \div \frac{x^2 - 4y^2}{x^2 + 2xy + y^2} \quad \left(= \frac{x^2 - 2xy}{x + y} \times \frac{x^2 + 2xy + y^2}{x^2 - 4y^2} \right)$$

$$= \frac{x(x - 2y)}{x + y} \times \frac{(x + y)(x + y)}{(x + 2y)(x - 2y)}$$

Now cancel common factors.

$$= \frac{x(x + y)}{(x + 2y)}$$

Exercise 2

1 Simplify these algebraic expressions and give the answers in their lowest terms.

a) $\dfrac{5xy^2}{4x^2} \times \dfrac{8x^3y}{15y^5}$

b) $\dfrac{a^2 - b^2}{a^4b} \times \dfrac{ab^2}{3a + 3b}$

c) $\dfrac{3a^2b - ab^2}{6a} \times \dfrac{9a^2}{9a^2 - b^2}$

d) $\dfrac{25 - x^2}{14x^3y^8} \times \dfrac{7x^2y}{8x + 40}$

e) $\dfrac{13xy^2}{x^2 + 3x - 18} \times \dfrac{x^2 - 9}{26x^4y^2}$

f) $\dfrac{-a^3 + ab^2}{a^2} \times \dfrac{a^3 + 7a^2b}{a^2 + 6ab - 7b^2}$

g) $\dfrac{x^2 - 3x - 10}{x + 7} \times \dfrac{3x + 21}{6x - 30}$

h) $\dfrac{a^2 + 5ab + 6b^2}{a^2 - 5ab + 6b^2} \times \dfrac{10a - 30b}{5a + 10b}$

i) $\dfrac{2x + 10}{32 - 8x} \times \dfrac{x^2 - 10x + 24}{x^2 - x - 30}$

2 Simplify these algebraic expressions and give the answers in their lowest terms.

a) $\dfrac{12m^2n^5}{m + 5} \div \dfrac{3m^3n}{m^2 - 25}$

b) $\dfrac{16 - 2m}{m^2 + 2m - 24} \div \dfrac{m - 8}{3m + 18}$

c) $\dfrac{m^2 - n^2}{m^2 + 2mn + n^2} \div \dfrac{m^2n - mn^2}{7m^2}$

d) $\dfrac{n^2 - n - 12}{2n^2 - 15n + 18} \div \dfrac{3n^2 - 12n}{2n^3 - 9n^2}$

e) $\dfrac{27a^4b^7}{3a^2 - 6a + 3} \div \dfrac{9ab^3}{(a - 1)^3}$

f) $\dfrac{a^2}{a^2 - 4} \div \dfrac{3a - a^2}{a^2 - 5a + 6}$

g) $\dfrac{9x}{x^2 - 25} \times \dfrac{x^2 + 5x}{2x - 4} \times \dfrac{x^2 + 3x - 10}{3x^4}$

h) $\dfrac{4x^2 - y^2}{x^2y - xy^2} \times \dfrac{x^2 + xy}{8x + 4y} \div \dfrac{2x^2 - 7xy + 3y^2}{8x^5y}$

i) $\dfrac{x^4 - y^4}{3x^2y - 3xy^2} \div \dfrac{x^2 + 2xy + y^2}{9xy^3} \div \dfrac{4x^2 + 4y^2}{xy^2 + y^3}$

j) $\dfrac{x^2 + 4xy + 3y^2}{x + 3y} \div \dfrac{x^2 - y^2}{x^2 - 2xy + y^2} \div \dfrac{x - y}{x + y}$

C Addition and subtraction of algebraic fractions

To add and subtract fractions, we have already learned that they must first be written as equivalent fractions all with the same denominator. This new denominator is the lowest common multiple (LCM) of the denominators of the fractions we are adding and subtracting.

We follow exactly the same method to add and subtract algebraic fractions. It is important that we factorise the expressions in the fractions first. This will help us to find the **lowest** common multiple, rather than simply multiplying the expressions in all the denominators to use as a common multiple.

Examples

Simplify these algebraic fractions and write the answers in their lowest terms.

a) $\dfrac{x + 4}{3} + \dfrac{x - 2}{5}$

First factorise the expressions if possible (not possible in this case). Then find the LCM.

The LCM of 3 and 5 is 15.

$$\dfrac{x + 4}{3} + \dfrac{x - 2}{5} = \dfrac{5(x + 4) + 3(x - 2)}{15}$$

Simplify the numerator.

$$= \dfrac{5x + 20 + 3x - 6}{15}$$

$$= \dfrac{8x + 14}{15}$$

b) $\dfrac{2}{3y} + \dfrac{1}{y^2}$

It is not possible to factorise the expressions. Find the LCM.

The LCM of $3y$ and y^2 is $3y^2$.

$$\frac{2}{3y} + \frac{1}{y^2} = \frac{(y \times 2) + (3 \times 1)}{3y^2}$$

Simplify the numerator.

$$= \frac{2y + 3}{3y^2}$$

c) $\dfrac{5}{2x + 2y} - \dfrac{4}{3x + 3y}$

First factorise the expressions if possible.

$$\frac{5}{2x + 2y} - \frac{4}{3x + 3y} = \frac{5}{2(x + y)} - \frac{4}{3(x + y)}$$

Then find the LCM.

The LCM of $2(x + y)$ and $3(x + y)$ is $6(x + y)$.

$$= \frac{(3 \times 5) - (2 \times 4)}{6(x + y)}$$

Simplify the numerator.

$$= \frac{15 - 8}{6(x + y)}$$

$$= \frac{7}{6(x + y)}$$

d) $\dfrac{x + 2}{x^2 + x - 2} + \dfrac{3}{x^2 - 1}$

First factorise the expressions if possible.

$$\frac{x + 2}{x^2 + x - 2} + \frac{3}{x^2 - 1} = \frac{x + 2}{(x + 2)(x - 1)} + \frac{3}{(x - 1)(x + 1)}$$

Cancel:

$$= \frac{1}{(x - 1)} + \frac{3}{(x - 1)(x + 1)}$$

The LCM of $(x - 1)$ and $(x - 1)(x + 1)$ is $(x - 1)(x + 1)$.

$$= \frac{(x + 1) + 3}{(x - 1)(x + 1)}$$

Simplify the numerator.

$$= \frac{(x + 4)}{(x - 1)(x + 1)}$$

e) $\dfrac{3}{4x} - \dfrac{4x + 12}{x^2 + 6x + 9} + \dfrac{15x}{5x^2 + 15x}$

Factorise the expressions (and cancel) if possible.

$$\frac{3}{4x} - \frac{4x + 12}{x^2 + 6x + 9} + \frac{15x}{5x^2 + 15x}$$

$$= \frac{3}{4x} - \frac{4(x + 3)}{(x + 3)^2} + \frac{15x}{5x(x + 3)} = \frac{3}{4x} - \frac{4}{(x + 3)} + \frac{3}{(x + 3)}$$

Then find the LCM.

The LCM of $4x$ and $(x + 3)$ is $4x(x + 3)$.

$$= \frac{3(x + 3) - 16x + 12x}{4x(x + 3)}$$

Simplify the numerator.

$$= \frac{3x + 9 - 4x}{4x(x + 3)}$$

$$= \frac{9 - x}{4x(x + 3)}$$

1 Simplify these algebraic expressions and give the answers in their lowest terms.

a) $\dfrac{2a}{3} + \dfrac{a-4}{2}$

b) $\dfrac{2m+1}{3} + \dfrac{m-3}{4}$

c) $\dfrac{3+2m}{4} - \dfrac{5}{9}$

d) $\dfrac{4n-5}{12} - \dfrac{6n+7}{16}$

e) $\dfrac{3}{2x^2} + \dfrac{7}{6x}$

f) $\dfrac{-4}{x^3} + \dfrac{9}{x} + \dfrac{2}{x^2}$

g) $\dfrac{1}{2} - \dfrac{3}{x} - \dfrac{5}{x^2}$

h) $3x - \dfrac{3}{x}$

i) $\dfrac{a+4}{3a} + \dfrac{2a-1}{5a^2}$

j) $\dfrac{9}{25a^3} - \dfrac{7}{15a}$

k) $\dfrac{13}{60h^2k^2} - \dfrac{11}{90hk^4}$

l) $\dfrac{a-4}{6a} + \dfrac{1}{3a^2} + \dfrac{7-3a}{9a^3}$

2 Simplify these algebraic expressions and give the answers in their lowest terms.

a) $\dfrac{8}{a+4} + 3$

b) $\dfrac{7}{3a-1} - 2$

c) $\dfrac{10}{x-3} - \dfrac{10}{x+5} + 2$

d) $\dfrac{8}{x^2-4} - \dfrac{3}{x-2}$

e) $\dfrac{5}{x^2-9} + \dfrac{2}{x-3} + 1$

f) $\dfrac{x}{x+5} + \dfrac{7x+10}{x^2+5x}$

g) $\dfrac{d+2}{4d-1} - \dfrac{7}{d+5}$

h) $\dfrac{m}{m+5} + \dfrac{10m}{m^2-25}$

i) $\dfrac{d^2+3}{d^2-2d} - \dfrac{d-4}{d}$

j) $\dfrac{9}{x^2-2x-15} - \dfrac{2}{x+3}$

k) $\dfrac{2}{m+3} + \dfrac{9}{m^2+8m+15}$

l) $\dfrac{2}{a-3} + \dfrac{7}{a^2+a-12} + \dfrac{1}{a+4}$

m) $\dfrac{d^2-11}{d^2-7d+12} - \dfrac{d+1}{d-4}$

n) $\dfrac{2x-5}{2x^2-16x+32} + \dfrac{4x+7}{x^2+x-20}$

o) $\dfrac{5}{2y} - \dfrac{3y-9}{y^2-6y+9} + \dfrac{12y}{4y^2-12y}$

D Solving equations that contain algebraic fractions

When solving an equation that contains algebraic fractions, first 'move' all the variables from the denominator so that they become part of the numerator.

The simplest way is to multiple both sides of the equation by the expression in the denominator. For example:

$$\frac{5}{x} = \frac{2}{x-3}$$

To move the 'x' in the denominator on the left, multiply both sides by x:

$$\frac{5}{x} \times x = \frac{2}{x-3} \times x \qquad \text{giving} \qquad 5 = \frac{2x}{x-3}$$

Now move the '$x - 3$' in the denominator on the right by multiplying both sides by $(x - 3)$:

$$5 \times (x-3) = \frac{2x}{x-3} \times (x-3) \qquad \text{giving} \qquad 5x - 15 = 2x$$

Now it is easy to solve for x as usual to give $x = 5$.

Examples

Solve each of these equations for the unknown variable.

a) $\dfrac{6}{2x-5} - \dfrac{4}{x-3} = 0$

If there is more than one algebraic fraction on the same side of the equation, use the methods we have learned to combine them over a single denominator.

$$\frac{6}{2x-5} - \frac{4}{x-3} = 0$$

$$\frac{6(x-3) - 4(2x-5)}{(2x-5)(x-3)} = 0$$

Multiply both sides by the expression in the denominator.

$$6(x-3) - 4(2x-5) = 0$$
$$6x - 18 - 8x + 20 = 0$$
$$-2x + 2 = 0$$
$$x = 1$$

b) $\dfrac{2}{y+3} - \dfrac{y-6}{y^2-9} = 0$

$\dfrac{2}{y+3} - \dfrac{y-6}{(y+3)(y-3)} = 0$

$\dfrac{2(y-3) - (y-6)}{(y+3)(y-3)} = 0$

Multiply both sides by the expression in the denominator.

$2(y-3) - (y-6) = 0$

$2y - 6 - y + 6 = 0$

$y = 0$

Exercise 4

1 Solve each of these equations for the unknown variable.

a) $\dfrac{5}{x} = \dfrac{6}{7}$

b) $\dfrac{4}{3g} + \dfrac{2}{6g} = 1\dfrac{1}{2}$

c) $\dfrac{3}{x-2} = \dfrac{1}{2}$

d) $\dfrac{7}{k} = \dfrac{5}{k-2}$

e) $x + 3 = \dfrac{6}{x+4}$

f) $\dfrac{7}{2x-1} = \dfrac{3}{x-4}$

g) $\dfrac{1}{v-2} = \dfrac{2}{v-1}$

h) $\dfrac{4}{u+3} - \dfrac{3}{u+2} = 0$

i) $\dfrac{5}{7x-6} - \dfrac{3}{5x+7} = 0$

j) $\dfrac{2a-3}{2} - \dfrac{a+2}{3} = \dfrac{a+1}{4}$

k) $\dfrac{3}{x+1} - \dfrac{1}{2x+2} = 5$

l) $\dfrac{1}{x+2} + \dfrac{3}{x+4} = \dfrac{4}{x+3}$

m) $\dfrac{5}{2x-1} - \dfrac{4}{4x-2} - \dfrac{3}{6x-3} = 1$

n) $\dfrac{3}{x-4} - \dfrac{x-5}{(x-4)(2x+3)} = \dfrac{4}{2x+3}$

o) $\dfrac{4}{m^2+3m+2} - \dfrac{3}{m^2+5m+6} = 0$

p) $\dfrac{3}{y-4} - \dfrac{y+2}{y^2-3y-4} = \dfrac{1}{2y+2}$

Unit 5 Simultaneous equations

So far, we have learned how to solve equations that have one variable. To solve an equation with one variable, we need only one equation that is true for that variable.

For example, $2x + 3 = 1$ is easy to solve, giving $x = -1$.

If the equation has two variables, we need more information to solve it. In fact we need two equations that are both true for the same variables. In general, to solve equations, we need the same number of different equations as the number of variables.

The word simultaneous means 'at the same time' and so simultaneous equations are ones that are true **at the same time** and for the **same values** of the variables.

Now we will look at pairs of different equations with **two variables**. We can solve these simultaneous equations using algebra or using graphs.

A Solving simultaneous equations with algebra

There are three basic algebraic methods that we can use to solve simultaneous equations:

1 Equating expressions
2 Elimination (addition or subtraction)
3 Substitution

We will learn all three methods, but normally you will choose the method that you think is most suitable for each pair of equations that you want to solve.

Equating expressions to solve simultaneous equations

Sometimes it is easy to write each of the equations so that the same term (containing only one of the variables) is on one side. The terms on the other side of each equation will then be equal, and we can solve for the other variable.

Example

Solve this pair of simultaneous equations.
$x + y = 7$ and $x - y = 3$

Rearrange each equation so that the like term 'x' is on the LHS.
$$x + y = 7 \quad \text{gives} \quad x = 7 - y$$
and $x - y = 3$ gives $x = 3 + y$
We know that the value of the variable x is the same in both equations, so we can say that:
$7 - y = 3 + y$
and we can now solve for the variable y:
$4 = 2y$
$y = 2$
Now substitute this value of $y = 2$ into the original two equations to find the value of x.
$x + y = 7$ gives $x + 2 = 7$, so $x = 5$
$x - y = 3$ gives $x - 2 = 3$, so $x = 5$
Both equations give the same value for x, which checks our answer. The solution to the equations is $x = 5$ and $y = 2$, which we can also write as $(5, 2)$.

Sometimes the coefficient of x is not 1.

Example

Solve the simultaneous equations $2x - 5y = -1$ and $2x = 3y + 1$.

Rearrange the equations.

$2x - 5y = -1$ gives $2x = 5y - 1$
But we also have $2x = 3y + 1$
So $5y - 1 = 3y + 1$
                          $2y = 2$
                          $y = 1$

Substitute $y = 1$ into the original equations.
          $2x - 5(1) = -1$
                    $2x = 4$, which gives $x = 2$

$2x = 3(1) + 1$, which also gives $x = 2$
The solution to these equations is $x = 2$, $y = 1$ or $(2, 1)$.

Exercise 1

1 Solve each of these pairs of simultaneous equations by equating expressions.

a) $x = 3y - 1$
$x = 2y + 1$

b) $4x + y = 9$
$2x - y = 3$

c) $3x + 2y = 13$
$3x - 2y = 5$

d) $3y = 2x + 1$
$3y = 3x + 4$

e) $2x + 5y = 13$
$2x + y = 9$

f) $3x + 3y = 15$
$3x - 5y = -41$

g) $5y - 3x = 2$
$5y = 8x - 3$

h) $3x + 4y = -8$
$x + 4y = 4$

i) $5y - 7 = 3x$
$5y + 7 = 17x$

j) $4x - y = 7$
$4x + 3y = 11$

k) $3x - y = 11$
$3x + 2y = -4$

l) $3y + 2 = 4x$
$3y - 2 = 2x$

Using the elimination method to solve simultaneous equations

In this method, we eliminate (remove) one of the variables in both equations. This is possible if we add or subtract the two equations, or some multiple of the equations.

We need the coefficient for one of the like terms (x or y) to be exactly the same in both equations. The sign of this term can be different in the two equations, as long as the coefficients are equal.

Example

Solve these simultaneous equations by using the elimination method.
$3x - y = 12$ and $2x + y = 13$

The coefficient of the y term in both equations is the same (although they have opposite signs).
If we **add** the two equations, we lose the y terms ($-y + y = 0$).

This gives
$$3x - y = 12$$
$$+(2x + y = 13)$$
$$\overline{5x \quad\quad = 25}$$
$$x = 5$$

Substitute $x = 5$ into the original equations:
$$3(5) - y = 12 \quad \text{and} \quad 2(5) + y = 13$$
So $\quad\quad\quad y = 3 \quad \text{and} \quad\quad\quad\quad y = 3$
The solution to these equations is $x = 5$, $y = 3$ or $(5, 3)$.

Example

Solve these simultaneous equations by using the elimination method.
$2x + 3y = 9$ and $2x + y = 7$

The coefficient of the x term in both equations is the same (2).
If we **subtract** the two equations, we lose the x terms.

This gives
$$\begin{aligned} 2x + 3y &= 9 \\ -(2x + y &= 7) \\ \hline 2y &= 2 \\ y &= 1 \end{aligned}$$

Substitute $y = 1$ into the original equations:
$2x + 3(1) = 9$ and $2x + 1 = 7$
So $x = 3$ and $x = 3$
The solution to these equations is $x = 3$, $y = 1$ or $(3, 1)$.

Sometimes the coefficients of neither x nor y terms are the same.

Example

Solve these simultaneous equations by using the elimination method.
$2x + y = -1$ and $3x - 4y = 4$

The coefficients of neither of the pairs of like terms are equal. But we can **make** two of them equal if we multiply the first equation by 4.

$4 \times (2x + y) = 4 \times -1$

This gives us these two new simultaneous equations:
$8x + 4y = -4$ and $3x - 4y = 4$

If we **add** the two equations, we lose the y terms.

This gives
$$\begin{aligned} 8x + 4y &= -4 \\ +(3x - 4y &= 4) \\ \hline 11x &= 0 \text{ or } x = 0 \end{aligned}$$

Substitute $x = 0$ into the original equations:
$2(0) + y = -1$ and $2(0) - 4y = 4$
So $y = -1$ and $y = -1$
The solution to these equations is $x = 0$, $y = -1$ or $(0, -1)$.

Sometimes we need to multiply both equations to obtain the LCM of the coefficients of like terms.

Example

Solve these simultaneous equations by using the elimination method.
$6x + 5y = 0$ and $4x - 3y = 38$

The coefficients of neither of the pairs of like terms are equal.
We can make the coefficients of the x terms equal (the LCM of 6 and 4),
or we can make the coefficients of the y terms equal (the LCM of 5 and 3).
We will choose to make the x terms equal. The LCM of 6 and 4 is 12, so we must multiply the first equation by 2 and the second equation by 3 to give $12x$ in each equation.

$2 \times (6x + 5y) = 2 \times 0$ and $3 \times (4x - 3y) = 3 \times 38$
So we have
$$12x + 10y = 0 \quad \text{and} \quad 12x - 9y = 114$$
If we **subtract** the two equations, we lose the x terms.

This gives
$$\begin{array}{r} 12x + 10y = 0 \\ -(12x - 9y = 114) \\ \hline 19y = -114 \text{ or } y = -6 \end{array}$$

Substitute $y = -6$ into the original equations:
$6x + 5(-6) = 0$ and $4x - 3(-6) = 38$
So $\quad x = 5$ and $\quad x = 5$
The solution to these equations is $x = 5$, $y = -6$ or $(5, -6)$.

Exercise 2

1 Solve each of these pairs of simultaneous equations by the elimination method.

a) $x - y = 5$
$\quad x + y = 19$

b) $2x - y = 2$
$\quad x + y = 7$

c) $x + y = 3$
$\quad x - y = 1$

d) $2x + y = 23$
$\quad 4x - y = 19$

e) $3x + y = 9$
$\quad 2x + y = 7$

f) $2x + y = 8$
$\quad 3x - y = 17$

g) $2x + 3y = 10$
$\quad 3y - x = 7$

h) $x + 4y = 11$
$\quad x + y = 5$

i) $6x - y = 23$
$\quad 3y + 6x = 11$

j) $2x - y = 10$
$\quad 3x + y = 10$

k) $3x - y + 14 = 0$
$\quad 2x + y + 1 = 0$

l) $x - 2y = 5$
$\quad 3x + 2y = -9$

m) $5x + 7y - 17 = 0$
$\quad 7y + 3x - 27 = 0$

n) $2x + y = 4$
$\quad 4x - y = 11$

o) $x + 3y = 9$
$\quad 2x - 3y = 0$

p) $3x + y = 17$
$\quad 3x - y = 19$

2 Solve each of these pairs of simultaneous equations by the elimination method.

a) $5x - 2y = 11$
   $x + y = 5$

b) $3x + y = 9$
   $x - 2y = 10$

c) $x + 3y = 38$
   $3x - y = 24$

d) $2x + 7y = 26$
   $x + y = 3$

e) $2x + y = 10$
   $-x + 2y = 9$

f) $3x - 2y = 6$
   $6y = 5x + 30$

g) $2x - 3y = 3$
   $3x + y = -23$

h) $3x + 4y = 6$
   $3y = 7 - x$

i) $3x + 7y = 1$
   $2x - 3y = 16$

j) $4x + 3y = 0$
   $7x - 2y = -29$

k) $-3x + 2y = 5$
   $4x + 3y = -1$

l) $6x - 12y = 0$
   $4x - 2y = 4\frac{1}{2}$

m) $5x - 4y = -1$
   $2x - 3y = 1$

n) $2x + 3y = 14$
   $8x - 5y = 5$

o) $x + \frac{1}{2}y = 4$
   $\frac{1}{2}x - \frac{1}{2}y = \frac{1}{2}$

p) $4x - 7y = 15$
   $5x - 12 = 2y$

Using the substitution method to solve simultaneous equations

When we **substitute**, we replace one thing by another one. To substitute for x, we rewrite one of the equations so that it reads '$x = \ldots$'. Then we use this expression for x to replace the variable x in the second equation. The second equation then contains only the variable y, so we can solve for y and then find x as before.

Similarly, to substitute for y, we rewrite one of the equations so that it reads '$y = \ldots$'. Then we use this expression for y to replace the variable y in the second equation, solve for x then find y as before.

Examples

a) Solve these simultaneous equations by using the substitution method.
   $2x + 3y = -5$ and $x = y - 10$

The second equation is already in the form '$x = \ldots$'.
We can substitute the expression $y - 10$ for x in the first equation.
$2(y - 10) + 3y = -5$
$\quad 2y - 20 + 3y = -5$
$\qquad\qquad 5y = 15$ so $y = 3$
Now substitute this value for y into one (or both to check) of the original equations.
$x = 3 - 10 = -7$
So the solution to these equations is $x = -7$ and $y = 3$ or $(-7, 3)$.

b) Solve these simultaneous equations by using the substitution method.
$10x + 7y = -1$ and $2x + y = 5$

We can rearrange the second equation to give $y = 5 - 2x$.
Now substitute the expression $5 - 2x$ for y in the first equation.

$10x + 7(5 - 2x) = -1$
$10x + 35 - 14x = -1$
$\qquad\qquad -4x = -36$
$\qquad\qquad\quad x = 9$

Now substitute this value for x into one (or both to check) of the original equations.
$2(9) + y = 5$
$\qquad\quad y = -13$
The solution to these equations is $x = 9$ and $y = -13$ or $(9, -13)$.

c) Solve these simultaneous equations by using the substitution method.
$7x + 4y = 12$ and $2x - 3y = -38$

We can rearrange the second equation to give $2x = 3y - 38$.
This gives $x = \frac{3}{2}y - 19$ to substitute into the first equation.

$7\left(\frac{3}{2}y - 19\right) + 4y = 12$
$10\frac{1}{2}y - 133 + 4y = 12$
$\qquad\qquad\quad 14\frac{1}{2}y = 145$
$\qquad\qquad\qquad\quad y = 10$

Now substitute this value for y into one (or both to check) of the original equations.

$7x + 4(10) = 12$
$\qquad\quad 7x = -28$
$\qquad\quad\; x = -4$
The solution to these equations is $x = -4$ and $y = 10$ or $(-4, 10)$.

Exercise 3

1 Solve each of these pairs of simultaneous equations by the substitution method.

a) $3x - y = 9$
  $y = 2x$

b) $3x + 2y = 4$
  $x = y - 2$

c) $3x + 4y = 11$
  $y = 9 - 2x$

d) $5x - 2y = 23$
  $x = y + 1$

e) $5x + 2y = -1$
  $x - 3y = -7$

f) $x + y = 7$
  $x - y = 5$

g) $x + y = -1$
  $3x - 4y = 4$

h) $2x + y = 17$
  $y - 6x = 1$

i) $3x - y = 0$
  $2x + y = 5$

j) $x - 10y = 23$
  $3x - 5y = 19$

k) $x + 5y = 13$
  $\frac{1}{3}x - y = 3$

l) $x - 2y = -3$
  $x + y = 3$

2 Solve each of these pairs of simultaneous equations by the substitution method.

a) $3x - 5y = 6$
$2x - 3y = 5$

b) $3x - 5y = 19$
$5x + 2y = 11$

c) $3x + 7y = 2$
$6x - 5y = 4$

d) $3x + 2y = 13$
$5x - 4y = 18$

3 Solve each of these pairs of simultaneous equations.
Choose any method you have learned that you think will be suitable.

a) $3x - y = 2$
$x + 2y = 17$

b) $2x + 3y = 9$
$x + 4y = 7$

c) $3x - 2y = 0$
$2x + 3y = 13$

d) $2x + 3y = 9$
$4x - y = 4$

e) $3x - 4y = 2$
$2x - 6y = -7$

f) $3x + 2y = 5$
$2x - 3y = 12$

g) $8x + 3y = 2$
$5x = 1 - 2y$

h) $4x + 3y = 9$
$2x + 5y = -13$

B Using graphs to solve simultaneous linear equations

In Unit 1, we learned about graphs of linear equations in two variables. In the first example on page 76, we used the elimination method to solve the two simultaneous linear equations $2x + 3y = 9$ and $2x + y = 7$. We will now draw the graph of each of these linear equations.

You will see that the two graphs intersect (cross) at one point: $(3, 1)$.
So the graphs tell us that the point $(3, 1)$ is on **both** the lines.

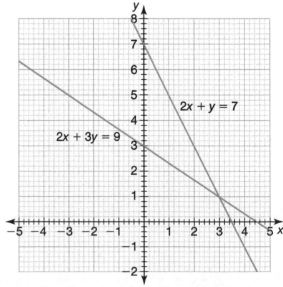

Solving the equations, we found that the values $x = 3$ and $y = 1$ are true for **both** the equations simultaneously.

So the point $(3, 1)$ is where the graphs intersect and it also gives the values of x and y that are true for both equations simultaneously. (You can check this is true by drawing the graphs of the equations in other examples.)

We can use this fact as another method to help us solve simultaneous linear equations.

Example

Solve these simultaneous equations by drawing their graphs.
$5x + 4y = 12$ and $x + 3y = -2$

Write the equations in the form $y = mx + c$ and choose points to plot.

$5x + 4y = 12$, so $y = -1\frac{1}{4}x + 3$

x	-2	-1	0	1	2
y	$5\frac{1}{2}$	$4\frac{1}{4}$	3	$1\frac{3}{4}$	$\frac{1}{2}$

$x + 3y = -2$, so $y = -\frac{1}{3}x - \frac{2}{3}$

x	-2	-1	0	1	2
y	0	$-\frac{1}{3}$	$-\frac{2}{3}$	-1	$-1\frac{1}{3}$

We see that the two graphs intersect at the point $(4, -2)$.
Check that these values are true for both equations.
First equation:
$5(4) + 4(-2) = 20 - 8 = 12$ ✓
Second equation: $4 + 3(-2) = 4 - 6 = -2$ ✓

The solution to these equations is $(4, -2)$.

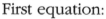

Exercise 4

1 Solve each of these pairs of simultaneous equations by drawing their graphs.

a) $3x - y = 0$
   $2x - y = 1$

b) $y = 3x - 1$
   $y = 2x + 1$

c) $3x + 2y = 12$
   $y = x + 1$

d) $y = 5$
   $3x + y = -4$

e) $3x - 2y = 7$
   $2x + 3y = 9$

f) $y = x - 2$
   $3x + y = 6$

g) $x + 3y = 6$
   $y = 2x - 5$

h) $3x + 2y = 4$
   $5x + y = 2$

i) $x + y = 4$
   $4x - y = 1$

j) $2x + 5y = 25$
   $3x - 2y = 9$

k) $3x + 4y = 24$
   $2y = x + 2$

l) $x - 2y = -1$
   $x - y = -2$

C Solving problems using simultaneous equations

Many problems that we meet in everyday life involve more than one unknown quantity. We can use simultaneous equations to find the answer to these problems.

Here is a reminder of the steps we met in Unit 3 that you can follow to help you solve word problems, here applied to simultaneous equations.

Step 1 Work in an organised manner. Read the whole problem first.
Step 2 Think about the problem. Write down what is given, and what you must find.
Step 3 Draw a diagram if appropriate; it helps to 'see' the problem.
Step 4 Choose variables (e.g. x and y) to stand in place of what you must find, or the values connected with what you must find.
Step 5 Change the information in the problem from words into maths sentences using symbols so that you can write equations. These equations should be simultaneously true for the variables.
Step 6 Solve the simultaneous equations to find the values of the unknown variables.
Step 7 Translate this result back into words to answer the question asked.
Step 8 Check the answer.

Example

Mary buys 2 skirts and 3 blouses. She pays $160.
Jane buys 3 of the same skirts and 2 of the same blouses. She pays $190.
Work out the price of a skirt and the price of a blouse.

Let the price of a skirt be $\$s$ and the price of a blouse be $\$b$.
We know: $2s + 3b = 160$ and $3s + 2b = 190$
Eliminate the b variable:

Multiply the first equation by 2: $\qquad 4s + 6b = 320$
Multiply the second equation by 3: $-(9s + 6b = 570)$
Then **subtract** $\qquad\qquad\qquad\qquad -5s \qquad = -250$
$\qquad\qquad\qquad\qquad\qquad\qquad\qquad s = 50$

Substitute $s = 50$ into an original equation:
$2(50) + 3b = 160$, so $b = 20$
The price of a skirt is $50 and the price of a blouse is $20.

Example

The sum of two numbers is 33 and their difference is 9.
Find the two numbers.

Let the bigger number be x and the smaller number be y.
We know: $x + y = 33$ and $x - y = 9$
Solve using substitution:
Rearrange the second equation $x - y = 9$, so $x = 9 + y$
Substitute this value for x into the first equation.

$(9 + y) + y = 33$
$2y = 24$
$y = 12$

Substitute this value for y into one of the original equations.
$x - 12 = 9$
$x = 21$
The two numbers are 12 and 21.

 Exercise 5

1 The sum of two numbers is 40 and their difference is 6.
   Find the two numbers.

2 Find two numbers so that twice the smaller number plus three times the
   larger one is 34, and five times the smaller number minus twice the larger
   one is 9.

3 The second of two numbers is 5 more than twice the first.
   The sum of the numbers is 44.
   Find the numbers.

4 The larger of two numbers is 8 more than four times the smaller. If the larger
   number is increased by four times the smaller one, the result is 40.
   Find the numbers.

5 The difference between two numbers is 16. Five times the smaller number is
   the same as 8 less than twice the larger one.
   Find the numbers.

6 The sum of two numbers is the same as four times the smaller number. If
   twice the larger number is decreased by the smaller one, the result is 30.
   Find the numbers.

7 The number of girls in Flowerville High School is 60 greater than the number
   of boys.
   If there are 1250 pupils in the school, how many girls are there?

8 Two groups of pupils go out for lunch. In the first group, two pupils have soup and five have sandwiches, and their bill is $8.00. In the second group, five pupils have soup and two have sandwiches, and their bill is $9.50. What is the price of soup?

9 The price of a sweater is $5 less than twice the price of a shirt. Four sweaters and three shirts cost $200.
Find the price of each shirt and each sweater.

10 A shop has 20 TV sets that weigh 880 kg together. Some of the TV sets weigh 30 kg each and the others weigh 50 kg each.
How many TV sets weigh 50 kg?

11 Five years ago, Mr Carr was three times as old as his son, Joe. Five years from now, Mr Carr's age will be twice his son's age.
How old are Mr Carr and Joe now?

12 The sides of an equilateral triangle measure $(x + y + 3)$ cm, $(3y - 2)$ cm and $(2x + y)$ cm.
Calculate the perimeter of the triangle.

Unit 6   The geometry of circles

In Unit 1 of Coursebook 2, we learned how to calculate the circumference and the area of a circle. You may want to revise these.

Revision

Circumference	the perimeter (length around the outside) of a circle
Radius	a line from the centre of the circle to any point on the circumference
Diameter	a line joining two points on the circumference and passing through the centre
Arc	part of the circumference of a circle

The **circumference** of a circle $= 2 \times \pi \times r$
where r is the radius of the circle.
The **area** of a circle $= \pi r^2$ where r is the radius of the circle.

Here are more words to describe other geometrical properties of circles:

Chord   a line joining any two points on the circumference

NOTE: The diameter is a special chord and it is also the longest chord.

Tangent   a line outside the circle that touches the circumference at one point only

Segment   a chord divides a circle into two segments

Sector   two radii (r) divide a circle into two sectors (a sector looks like a slice of pie)

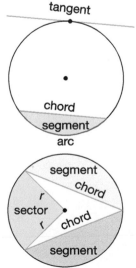

We know that every angle (except 180°) has two parts: the big or **major** angle and the small or **minor** angle.

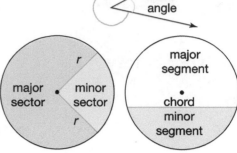

This is also true for sectors and segments. In a circle, the bigger segment (or sector) is called the **major segment** (or **sector**) and the smaller segment (or sector) is called the **minor segment** (or **sector**).

A Length of an arc and area of a sector

Look at these special sectors.

In the first diagram, the circle has been divided into four equal sectors.

So the **area** of one of these sectors
= $\frac{1}{4}$ × area of the whole circle.

Also, the length of the **arc** in each sector
= $\frac{1}{4}$ × circumference of the circle.

In the second diagram, the circle has been divided into two equal sectors.

So the **area** of one of these sectors
= $\frac{1}{2}$ × area of the whole circle.

Also, the length of the **arc** in each sector
= $\frac{1}{2}$ × circumference of the circle.

In the third diagram, the circle has been divided into three equal sectors.

So the **area** of one of these sectors
= $\frac{1}{3}$ × area of the whole circle.

Also, the length of the **arc** in each sector
= $\frac{1}{3}$ × circumference of the circle.

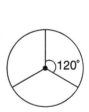

But we also know: $\dfrac{90°}{360°} = \dfrac{1}{4}$ and $\dfrac{180°}{360°} = \dfrac{1}{2}$ and $\dfrac{120°}{360°} = \dfrac{1}{3}$

In each case:

$$\text{area of the sector} = \frac{\text{angle of the sector}}{360°} \times \text{area of the circle}$$

and $$\text{length of arc in the sector} = \frac{\text{angle of the sector}}{360°} \times \text{circumference}$$

This is true for **every** sector.

For any sector of a circle with an angle of $x°$:

the length of its arc $= \frac{x°}{360°} \times$ circumference $= \frac{x°}{360°} \times 2\pi r$

the area of the sector $= \frac{x°}{360°} \times$ area $= \frac{x°}{360°} \times \pi r^2$

Examples

a) Calculate the area and the perimeter of the **minor** sector in this circle. Use $\pi = \frac{22}{7}$.

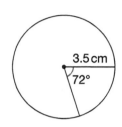

Area of sector $= \frac{72°}{360°} \times \pi r^2$

$= \frac{72}{360} \times \frac{22}{7} \times \frac{7}{2} \times \frac{7}{2}$

$= 7.7$

Area of minor sector is $7.7\,\text{cm}^2$.

Length of arc $= \frac{72°}{360°} \times 2\pi r$

$= \frac{72}{360} \times 2 \times \frac{22}{7} \times \frac{7}{2}$

$= 4.4$

Perimeter $= 4.4 + 3.5 + 3.5$

The perimeter of the minor sector is $11.4\,\text{cm}$.

b) Calculate the area and the perimeter of the **major** sector in this circle. Use $\pi = \frac{22}{7}$.

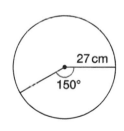

Area of sector $= \frac{360° - 150°}{360°} \times \pi r^2$

$= \frac{210}{360} \times \frac{22}{7} \times 27 \times 27$

$= 1336.5$

The area of the major sector is $1336.5\,\text{cm}^2$.

Length of arc $= \frac{360° - 150°}{360°} \times 2\pi r$

$= \frac{210}{360} \times 2 \times \frac{22}{7} \times 27$

$= 99$

Perimeter $= 99 + 27 + 27$

The perimeter of the major sector is $153\,\text{cm}$.

c) A circle has a radius of $24\,\text{cm}$. The arc of a sector of this circle has a length of $8\,\text{cm}$. Calculate the angle of this sector correct to the nearest whole number. Use $\pi = 3.142$.

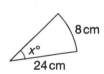

Length of arc $= \frac{x°}{360°} \times 2\pi r$

$8 = \frac{x}{360} \times 2 \times 3.142 \times 24$

$x = 19$

The angle of the sector is $19°$ (to nearest degree).

d) A circle has a radius of 8 mm. A sector of this circle has an area of 25.5 mm².
Calculate the angle of this sector correct to 1 decimal place. Use $\pi = 3.142$.

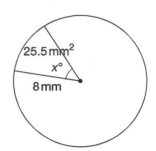

Area of sector $= \frac{x^{\circ}}{360^{\circ}} \times \pi r^2$

$$25.5 = \frac{x}{360} \times 3.142 \times 8 \times 8$$

$$x = 45.7$$

The angle of the sector is 45.7° (to 1 d.p.).

 Exercise 1

1 Copy and complete this table by calculating the missing values. Use $\pi = \frac{22}{7}$.

	Radius of circle	Diameter of circle	Circumference of circle	Area of circle	Angle of sector	Length of arc	Area of sector	Perimeter of sector
a)	7 cm				30°			
b)				616 m²			77 m²	
c)			220 mm			66 mm		
d)	28 cm							188 cm
e)					99°	24.2 m		
f)					324°		3465 mm²	
g)				$707\frac{1}{7}$ m²	75°			
h)		28 cm						80.8 cm

2 Copy and complete this table by calculating the missing values.
Use $\pi = 3.142$. Give your answers correct to 1 d.p.

	Radius of circle	Diameter of circle	Circumference of circle	Area of circle	Angle of sector	Length of arc	Area of sector	Perimeter of sector
a)		18 cm						41.6 cm
b)				50.3 mm²	50°			
c)					140°		175.9 cm²	
d)					324°	85 cm		
e)	24 m							58.6 m
f)			88.0 m			84.6 m		
g)				254.5 cm²			18.9 cm²	
h)	16 cm				100°			

Angle properties of circles

Starting with an arc *AB*, the chord *AB* can also be drawn between the same two points.

By joining these two points to the centre of the circle, *C*, we form the sector *ACB* and the angle ∠*ACB*.

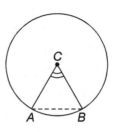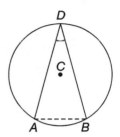

We can also join these two points to another point on the circumference (any other point) to make ∠*ADB*.

Now we have two angles 'standing' on the chord *AB* (or the arc *AB*): one is at the centre of the circle and the other is at the circumference.

NOTE: It is sometimes said that these angles are 'subtended' by the chord *AB*, or the arc *AB*, although 'standing' on the arc (or chord) is easier to understand.

If you construct and measure many different examples of these angles with different chords, you will find that, for any chord, the angle at the centre is exactly **twice** the size of the angle at the circumference.

Angle property 1
If two angles are standing on the same chord (or arc) and in the same segment, the angle at the centre is twice the angle at the circumference.

The diameter of a circle is a special chord that passes through the centre of the circle. The angle at the centre is always 180° (it is a straight line).

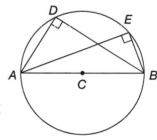

So any angle at the circumference that is standing on the diameter will be a right angle (90°).
∠*ADB* = ∠*AEB* = 90°

Angle property 2
Any angle at the circumference and standing on the diameter is equal to 90°, i.e. the angle in a semicircle is a right angle.

From Angle property 1, we know that any angle at the circumference standing on the same chord (in the same segment) is half the size of the angle at the centre standing on that chord.

The angle at the centre remains the same for that chord, so **all** the angles at the circumference (in the same segment) must be the same size.

$$\angle ADB = \angle AEB$$

(If the chord is the diameter, these angles will be 90°.)

> **Angle property 3**
> Angles at the circumference standing on the same chord (or arc) and in the same segment are equal.

Cyclic geometric figures

If all the vertices of a geometric figure lie on the circumference of one circle, we say this geometric figure is cyclic ('inside a circle').

In the diagrams at the top of page 89, we can say that $\triangle ADB$ is a cyclic triangle, but $\triangle ACB$ is not a cyclic triangle as one of its vertices (C) is not on the circumference.

A **cyclic quadrilateral** is a figure with four sides, and four angles, whose vertices are all on the circumference of a circle.

> **Angle property 4**
> The opposite angles in any cyclic quadrilateral are supplementary.
>
> (So $a + e = 180°$ and $b + d = 180°$.)

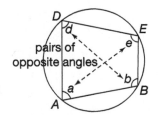

If we extend one of the sides of a cyclic quadrilateral outside the circle, it makes an angle that is outside the quadrilateral. This angle is called the exterior **angle** of the cyclic quadrilateral.

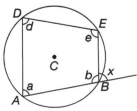

$$b + x = 180° \quad \text{(straight line)}$$
But also $b + d = 180°$ (angle property 4)
So $x = d$

> **Angle property 5**
> The exterior angle of any cyclic quadrilateral is equal to its opposite interior angle.

We can use all these angle properties to help us work out the size of angles in a circle.

Examples   Find the sizes of the unknown angles in each diagram.

a)

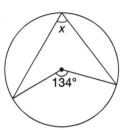

$x = \frac{1}{2} \times 134°$ (centre $\angle$ = 2 × circum. $\angle$)

$x = 67°$

b)

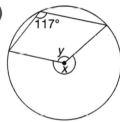

$x = 2 \times 117°$ (centre $\angle$ = 2 × circum. $\angle$)

$x = 234°$

$y = 360° - 234°$ ($\angle$s at point)

$y = 126°$

c)

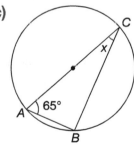

$\angle ABC = 90°$ ($\angle$ in semicircle)

$x = 180° - (90 + 65)°$ ($\angle$s in $\triangle$)

$x = 25°$

d)

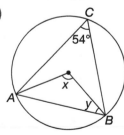

$x = 2 \times 54°$ (centre $\angle$ = 2 × circum. $\angle$)

$x = 108°$

$y = \frac{180° - 108°}{2} = 36°$ (base $\angle$s of isosceles $\triangle$)

Examples   Find the sizes of the unknown angles in each diagram.

a)

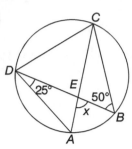

$\angle ACB = 25°$ (∠s in same segment)

$x = \angle ACB + \angle CBD$ (exterior ∠ of △ = sum of interior opp. ∠s)

$x = 25° + 50° = 75°$

b)

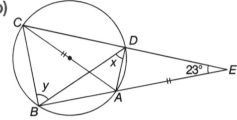

$\angle ACE = 23°$ (base ∠s of isosceles △)
$\angle ABC = 90°$ (∠ in semicircle)
$\angle ABD = 23°$ (∠s in same segment)
   $y = 90° - 23° = 67°$

$\angle BCE = 180° - (90° + 23°)$ (sum of ∠s in △BCE)
       $= 67°$

$\angle BCA = 67° - 23° = 44°$
   $x = \angle BCA = 44°$ (∠s in same segment)

c)

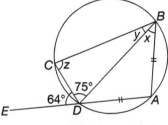

$\angle BDA = 180° - (64° + 75°)$ (∠s on straight line)
       $= 41°$

$x = \angle BDA = 41°$ (base ∠s of isosceles △)
$y = 64° - x$ (exterior ∠ of cyclic quad. = opposite interior ∠)

   $= 64° - 41° = 23°$

$z = 180° - (y + 75°)$ (sum of ∠s in △BCD)
   $= 180° - 98° = 82°$

NOTE: Remember that there are often many different 'routes' in geometry to find the angles.

Exercise 2

1 Calculate the size of each of the angles marked with a letter.

a)
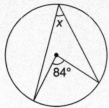

b)
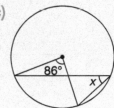

c)
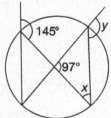

d)
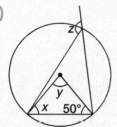

e)
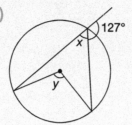

2 Calculate the size of each of the angles marked with a letter.

a)
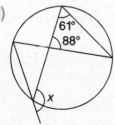

b)
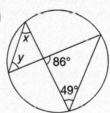

c)
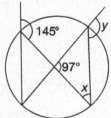

d)
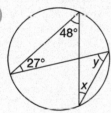

e)
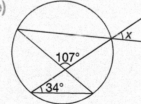

3 Calculate the size of each of the angles marked with a letter.

a)
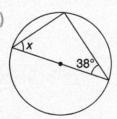

b)
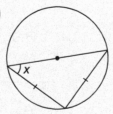

c)
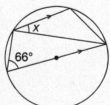

d)
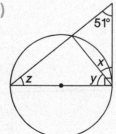

e)
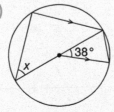

4 Calculate the size of each of the angles marked with a letter.

a)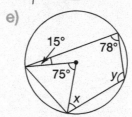

b)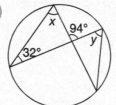

c)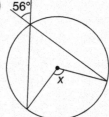

d)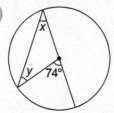

e)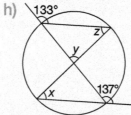

5 Calculate the size of each of the angles marked with a letter.

a)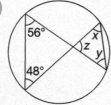

b)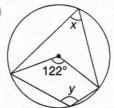

c)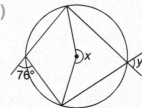

d)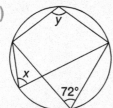

e)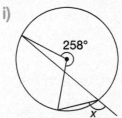

f)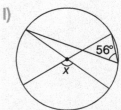

g)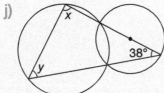

h)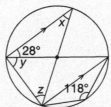

i)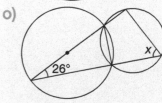

j)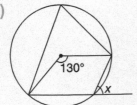

k)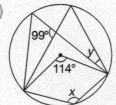

l)

m)

n)

o)

Tangents

A tangent is a straight line outside a circle that touches the circumference at one point only. If we join this point to the centre of the circle, the radius we draw is at right angles to the tangent.

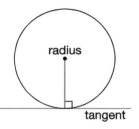

Tangent property 1
A tangent is perpendicular to the radius at the point of contact with the circumference.

If two tangents are drawn from the same point to touch a circle, the length of each tangent is the same.

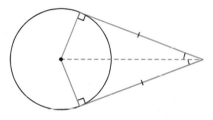

Tangent property 2
i) The two tangents drawn from the same point to a circle are equal.
ii) The line joining the point to the centre of the circle bisects the angle between the two tangents.

A chord divides a circle into two segments. If a tangent is drawn at one end of the chord, the chord and the tangent form a minor angle on the side of the minor segment, and a major angle on the side of the major segment.

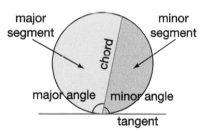

If an angle is standing on this chord, it has a special property.

1 The **minor angle** with the tangent is equal to the **angle in the major segment** standing on the chord.

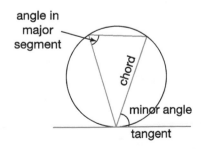

2 The **major angle** with the tangent is equal to the **angle in the minor segment** standing on the chord.

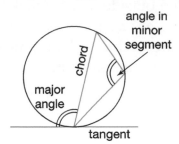

3 We can combine **1** and **2**.

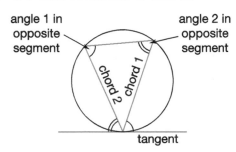

Tangent property 3

The angle between a tangent and a chord is equal to any angle in the **opposite segment** that is standing on this chord.

NOTE: This is sometimes called the 'alternate' segment, but 'opposite' segment is easier to understand.

Examples

Calculate the size of each angle marked with a letter.

a)

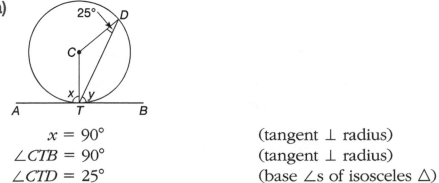

$$x = 90°$$ (tangent ⊥ radius)

$$\angle CTB = 90°$$ (tangent ⊥ radius)

$$\angle CTD = 25°$$ (base ∠s of isosceles △)

$$y = 90° - 25° = 65°$$

b)

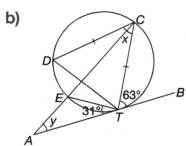

$\angle TDC = 63°$ (∠ in opp. segments)

$\angle DTC = 63°$ (base ∠s of isosceles △)

$x = 180° - (63° + 63°) = 54°$

$\angle ACT = 31°$ (∠ in opp. segments)

$\angle ATC = 180° - 63° = 117°$ (∠s on straight line)

$y = 180° - (117° + 31°)$ (sum of ∠s in △)

    $= 32°$

Exercise 3

1 Calculate the size of each angle marked with a letter.

a)

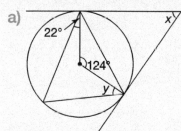

b)

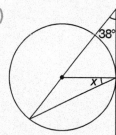

c)

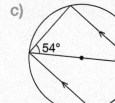

d)

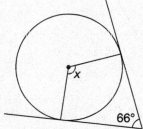

e)

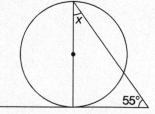

f)

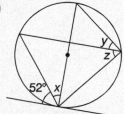

2 Calculate the size of each angle marked with a letter.

a)

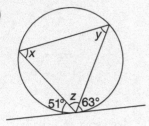

b)

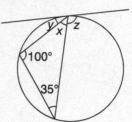

c)

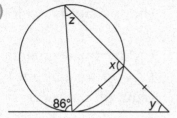

d)

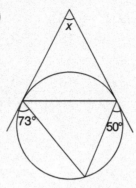

e)

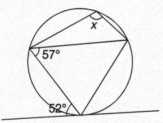

f)

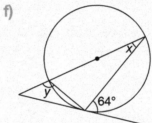

3 Calculate the size of each angle marked with a letter.

a)

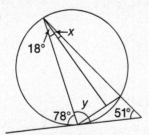

b)

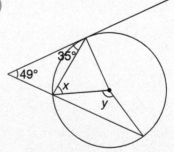

c)

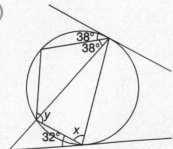

D Summary of circle properties

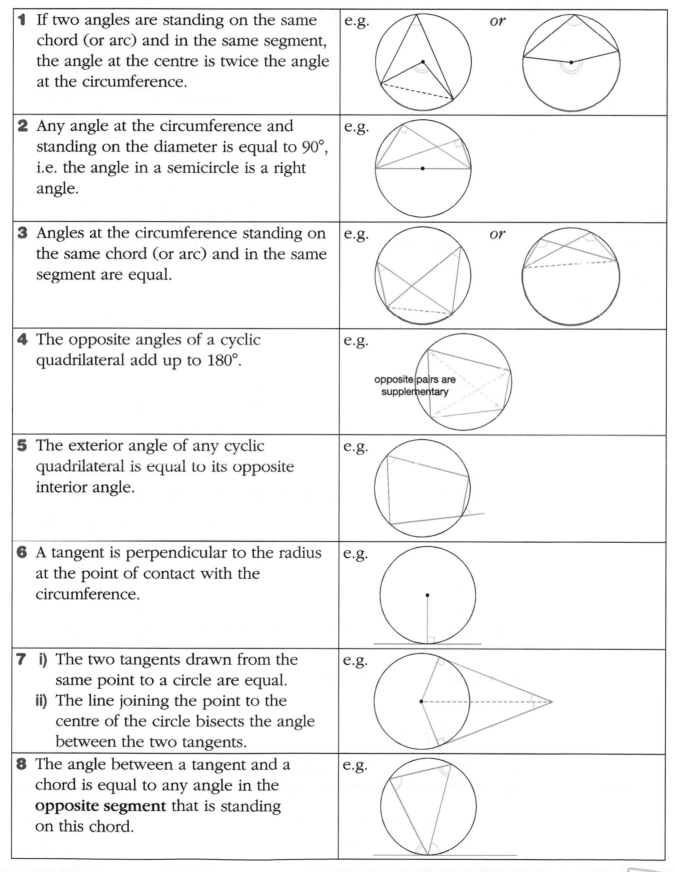

1 If two angles are standing on the same chord (or arc) and in the same segment, the angle at the centre is twice the angle at the circumference.	e.g. *or*
2 Any angle at the circumference and standing on the diameter is equal to 90°, i.e. the angle in a semicircle is a right angle.	e.g.
3 Angles at the circumference standing on the same chord (or arc) and in the same segment are equal.	e.g. *or*
4 The opposite angles of a cyclic quadrilateral add up to 180°.	e.g. opposite pairs are supplementary
5 The exterior angle of any cyclic quadrilateral is equal to its opposite interior angle.	e.g.
6 A tangent is perpendicular to the radius at the point of contact with the circumference.	e.g.
7 i) The two tangents drawn from the same point to a circle are equal. ii) The line joining the point to the centre of the circle bisects the angle between the two tangents.	e.g.
8 The angle between a tangent and a chord is equal to any angle in the **opposite segment** that is standing on this chord.	e.g.

Unit 7 Transformations

Key vocabulary

angle of rotation	isometric	rotation
centre of enlargement	line of reflection	scale factor
centre of rotation	map	transform
clockwise	mirror line	transformation
column vector	object	translation
enlargement	reduction	vector
image	reflect	
invariant point	reflection	

A **transformation** is an operation (something we do) that **transforms** (changes or moves) one point or a whole figure into another point or figure.

The starting point or figure before the transformation is called the **object**. The point or figure after the transformation is called the **image**.

1 Some transformations change only the **position** of the figure. The image is exactly the same size and shape as the object.
2 Some transformations keep only the **shape** of the object **the same**. The image has a different size. It can be smaller or bigger than the object.
3 Some transformations **change the shape** of the object so that the image looks different.

We will study only the kinds of transformations in **1** and **2** above.

A Translation

A **translation** is a transformation that moves all the points in a figure the **same distance** and in the **same direction**.

So a translation is just a movement along a straight line. The image is the same size and shape as the object but in a different place.

For example:

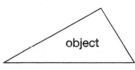 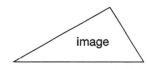

If we move this
triangle from here ...   ... to here, this is a **translation**.

We describe a translation by saying how far the figure moves
horizontally, and how far it moves **vertically**.

This is very similar to the way we describe the position of a point
on the Cartesian plane with **coordinates**. Coordinates tell us how
far the point is from the **origin** in the x-direction and also in the
y-direction (x, y).

We want to describe the movement on the Cartesian plane from **any**
point (not only the origin) to any other point. To do this, we use
column vectors (usually simply called **vectors**).
Vectors also have an 'x-part' and a 'y-part' that tell us how far to
move in the x-direction and the y-direction. So that we do not
confuse vectors with coordinates, for vectors we write the 'x-part'
above the 'y-part' inside brackets.

For example, the vector $\begin{pmatrix} 1 \\ 5 \end{pmatrix}$ tells us to move $+1$ in the x-direction

and $+5$ in the y-direction.

Coordinates describe the **position** of a point.
Vectors describe **movement**.

If we move the **object** (point A) so that it
is now at the place of the **image** (called
point A'), we have completed a
translation.

Now we must find the **vector** to describe
this translation.
The object has moved 3 places to the right.
This is an 'x' movement and it is positive.
The object has moved 4 places up.
This is a 'y' movement and it is also positive.

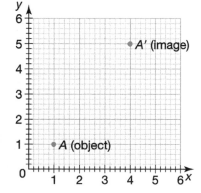

So, we write the vector for this translation as: $\begin{pmatrix} +3 \\ +4 \end{pmatrix}$

(If a number is positive, we don't always need to write the '+' sign.)

When we translate one point to another using a vector, we say that the object point is **mapped** onto the image point.

Movements in the *x*-direction
Movements to the **right** are shown with **positive** numbers (+).
Movements to the **left** are shown with **negative** numbers (−).

Movements in the *y*-direction
Movements **up** are shown with **positive** numbers (+).
Movements **down** are shown with **negative** numbers (−).

Sign of numbers in the vector	Movement	Sign of numbers in the vector	Movement
$\begin{pmatrix} + \\ + \end{pmatrix}$	to the **right** and **up**	$\begin{pmatrix} - \\ + \end{pmatrix}$	to the **left** and **up**
$\begin{pmatrix} + \\ - \end{pmatrix}$	to the **right** and **down**	$\begin{pmatrix} - \\ - \end{pmatrix}$	to the **left** and **down**

Example

An object *P* is at the point $(4, -6)$. It is translated along the vector $\begin{pmatrix} -4 \\ +7 \end{pmatrix}$.
Find the coordinates of the image P'.

The *x*-coordinate is moved 4 places to the **left**.
So it becomes $4 - 4 = 0$.
The *y*-coordinate is moved 7 places **up**.
So it becomes $-6 + 7 = +1$.
So the coordinates of the image P' are $(0, 1)$.

It is also possible to find this answer using the Cartesian plane.
Plot the point $P(4, -6)$.
Translate it 4 places to the left and 7 places up, to plot the image P'.
Read the coordinates of this image from the axes, $(0, 1)$.

NOTE: We show a column vector on the Cartesian plane using a straight line with an arrow in the middle pointing in the direction of the movement.

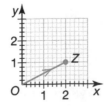

The vector shown here by the line *OZ* is $\begin{pmatrix} 2 \\ 1 \end{pmatrix}$.

Exercise 1

1 Write these translations using column vectors in the form $\begin{pmatrix} x \\ y \end{pmatrix}$.

a) 2 right, 3 up
d) 8 down

b) 12 left, 9 up
e) 4 right, 3 up

c) 2 right, 1 down
f) 5 left, 8 down

2 Describe in words the translations shown by these column vectors.
(Use 'left', 'right', 'up', 'down'.)

a) $\begin{pmatrix} 11 \\ 9 \end{pmatrix}$

b) $\begin{pmatrix} 2 \\ 1 \end{pmatrix}$

c) $\begin{pmatrix} -7 \\ 14 \end{pmatrix}$

d) $\begin{pmatrix} 6 \\ -2 \end{pmatrix}$

e) $\begin{pmatrix} -4 \\ -8 \end{pmatrix}$

f) $\begin{pmatrix} 0 \\ 3 \end{pmatrix}$

g) $\begin{pmatrix} 3.7 \\ 0 \end{pmatrix}$

h) $\begin{pmatrix} 0 \\ 0 \end{pmatrix}$

3 Use column vectors to describe the translations of these points as shown.

a)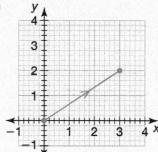

b)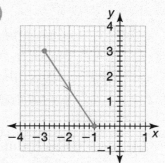

c)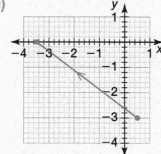

d)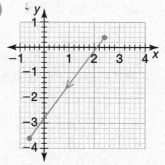

4 On graph paper, draw a Cartesian plane and plot the point $P(3, 3)$.
Translate point P to a new point using each of these vectors.
Give each new point a label and write down its coordinates.

a) $\begin{pmatrix} 1 \\ 1 \end{pmatrix}$

b) $\begin{pmatrix} 2 \\ -1 \end{pmatrix}$

c) $\begin{pmatrix} -3 \\ 2 \end{pmatrix}$

d) $\begin{pmatrix} -1.5 \\ -2 \end{pmatrix}$

e) $\begin{pmatrix} 3.5 \\ 2.5 \end{pmatrix}$

f) $\begin{pmatrix} 2 \\ 0 \end{pmatrix}$

g) $\begin{pmatrix} 0 \\ -3.5 \end{pmatrix}$

h) $\begin{pmatrix} 0 \\ 0 \end{pmatrix}$

5 Find the coordinates of the image of each of these points after translation
along the vector $\begin{pmatrix} 5 \\ -7 \end{pmatrix}$.

a) $(3, 6)$

b) $(-4, 7)$

c) $(-5, -6)$

d) $(8, -5)$

6 The point $(-6, 4)$ is mapped onto the point $(3, 7)$ along the translation vector, V. Find the coordinates of the image of each of these points after translation along the same vector, V.

a) $(-5, 6)$ b) $(7, -2)$ c) $(1, 1)$ d) $(6, 1)$

So far, we have moved a single point by translating it along a vector. How do we translate a whole figure (e.g. triangle, square) along a vector?

A line is made up of many points. So, to translate a whole line, we translate each point on the line along the same vector. In practice, we simply translate the **vertex points** of a figure and then join these points with straight lines.

> To **translate a figure** along a vector:
> - translate **each of the vertex points** along the vector
> - redraw the figure in the new position by joining the vertex points with straight lines.

Example

Translate this triangle ABC along the vector $\begin{pmatrix} 5 \\ 3 \end{pmatrix}$.

Find the coordinates of the vertices of the image triangle $A'B'C'$.

Each vertex point moves 5 places to the **right** and 3 places **up**. These moves are shown by the vector arrows.

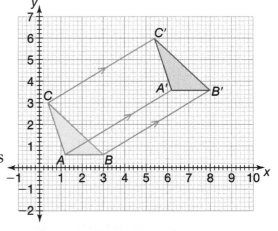

Plot the vertex points of the image triangle on the graph as shown and join them to draw the image triangle.
Read off the coordinates of the image vertex points.

$A(1.2, 0.6)$ moves to $A'(6.2, 3.6)$

$B(3, 0.6)$ moves to $B'(8, 3.6)$

$C(0.4, 3)$ moves to $C'(5.4, 6)$

Example

A rectangle R is translated along the vector $\begin{pmatrix} -6 \\ 2 \end{pmatrix}$.

Draw the image rectangle R' and work out the coordinates of its vertices.

Each vertex moves 6 places to the left and 2 place up. These moves are shown by the vector arrows.

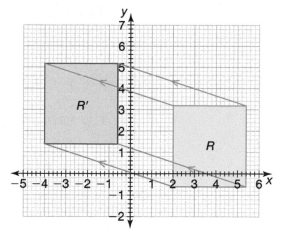

Plot the vertex points of the image rectangle on the graph as shown and join them to draw the image rectangle.

Read off the coordinates of the image vertex points:

$(-0.6, 1.4)$ $(-4, 1.4)$ $(-4, 5.2)$ and $(-0.6, 5.2)$

Examples

Write down the vector that describes each of the translations shown in diagram.

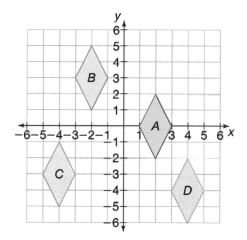

a) A onto B

Find the coordinates of the corresponding vertices on the two shapes. Remember to keep the order the same as the translation.

$(2, 2)$ is mapped onto $(-2, 5)$.
$2 + x = -2 \quad$ so $x = -4$
$2 + y = 5 \quad$ so $y = 3$

The vector is $\begin{pmatrix} -4 \\ 3 \end{pmatrix}$.

b) A onto C

$(2, -2)$ is mapped onto $(-4, -5)$.
$2 + x = -4 \quad$ so $x = -6$
$-2 + y = -5 \quad$ so $y = -3$

The vector is $\begin{pmatrix} -6 \\ -3 \end{pmatrix}$.

c) *A* onto *D*

(1, 0) is mapped onto 3, −4).
1 + *x* = 3   so *x* = 2
0 + *y* = −4   so *y* = −4

The vector is $\begin{pmatrix} 2 \\ -4 \end{pmatrix}$.

d) *C* onto *B*

(−3, −3) is mapped onto (−1, 3).
−3 + *x* = −1   so *x* = 2
−3 + *y* = 3   so *y* = 6

The vector is $\begin{pmatrix} 2 \\ 6 \end{pmatrix}$.

NOTE: We used only one vertex of each figure in the calculations. You can choose any of the vertices as long as you make sure they are corresponding vertices on the two figures. That means they are 'in the same place' on the figures. You can check your answer by choosing a second pair (or more) of corresponding vertices.

Example

The figure *P* is translated along the vector $\begin{pmatrix} -3 \\ 4 \end{pmatrix}$ to give the image *P'* shown in the diagram.

Find the coordinates of the object *P* and draw it on the same Cartesian plane.

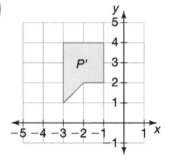

The vector $\begin{pmatrix} -3 \\ 4 \end{pmatrix}$ describes the movement of the figure *P* onto the image *P'*. So, to find the object *P*, we need to translate the coordinates of *P'* backwards along the same vector.

The reverse direction of the vector $\begin{pmatrix} -3 \\ 4 \end{pmatrix}$ is described as $\begin{pmatrix} +3 \\ -4 \end{pmatrix}$ so that each coordinate is moved the same distance but in the opposite direction.

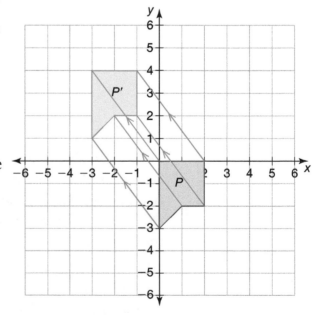

So   vertex (−1, 2) comes from the point (2, −2)
       vertex (−1, 4) comes from the point (2, 0)
       vertex (−3, 4) comes from the point (0, 0)
       vertex (−3, 1) comes from the point (0, −3)
and vertex (−2, 2) comes from the point (1, −2).

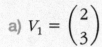 **Exercise 2**

1 Copy the figure *A* onto graph or squared paper.
Draw the image after translation along each of these vectors.

a) $V_1 = \begin{pmatrix} 2 \\ 3 \end{pmatrix}$ b) $V_2 = \begin{pmatrix} -1 \\ 2 \end{pmatrix}$

c) $V_3 = \begin{pmatrix} 3 \\ -5 \end{pmatrix}$ d) $V_4 = \begin{pmatrix} -4 \\ -6 \end{pmatrix}$

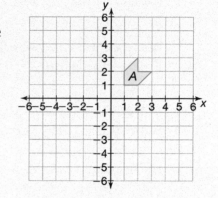

2 The diagram shows quadrilateral *ABCD*. Write down the coordinates of the image quadrilateral after *ABCD* has been translated along each of these vectors.

a) $\begin{pmatrix} 7 \\ 2 \end{pmatrix}$ b) $\begin{pmatrix} -10 \\ 6 \end{pmatrix}$

c) $\begin{pmatrix} 9 \\ -12 \end{pmatrix}$ d) $\begin{pmatrix} -11 \\ -8 \end{pmatrix}$

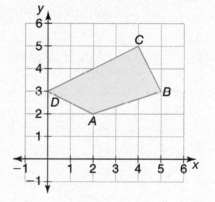

3 a) Write down the vector that will map the shaded triangle onto each of the labelled image triangles.
   b) Write down the vector used to map $\triangle E$ onto $\triangle C$.
   c) Write down the vector used to map $\triangle B$ onto $\triangle D$.
   d) Write down the vector used to map $\triangle D$ onto $\triangle A$.

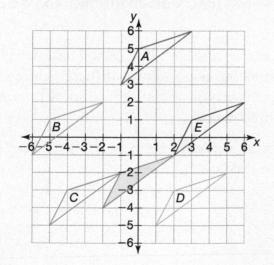

4 The diagram shows the image of the figure G after translation along the vector $\begin{pmatrix} 5 \\ -7 \end{pmatrix}$.

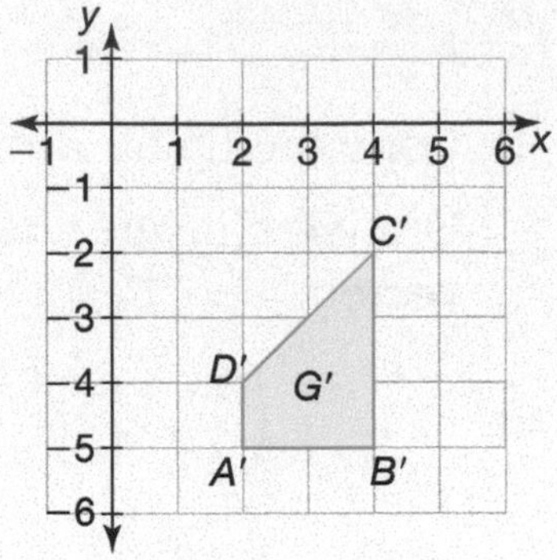

Write down the coordinates of the object figure G before translation.

5 On graph or squared paper draw the triangle XYZ with $X(1, 1)$, $Y(1, 4)$ and $Z(2, 1)$.

a) Translate $\triangle XYZ$ along the vector $\begin{pmatrix} 2 \\ 1 \end{pmatrix}$ and label this image $\triangle A$.

b) Translate $\triangle A$ along the vector $\begin{pmatrix} 1 \\ 1 \end{pmatrix}$ and label this image $\triangle B$.

c) Translate $\triangle B$ along the vector $\begin{pmatrix} 3 \\ -5 \end{pmatrix}$ and label this image $\triangle C$.

d) Work out the vector that would map $\triangle XYZ$ directly onto:
 i) $\triangle B$
 ii) $\triangle C$.

B Reflection

Looking in a mirror, we see a 'picture' that is an exact copy of our face. We call this exact copy of something in a mirror a **reflection**.

Our reflection appears to be the same distance behind the mirror as our face is in front of it. You will have studied this fact about mirrors in science lessons.

In maths, we talk about a **mirror line** or a **line of reflection**. This is a straight line that we choose to act like a mirror. If there is a figure on one side of the mirror line, we can draw the reflection of the figure on the other side of the mirror line.

We can choose any convenient straight line to be a mirror line. Often we choose the x-axis or the y-axis.

Drawing the reflection of a figure

We measure the **perpendicular distance** from a point to the mirror line. We know that the reflection of this point is the **same perpendicular distance** behind the mirror line. If we want to draw the reflection of a whole figure, we only need to **reflect** the vertex points in this way and then join them to draw the reflection of the figure, as we did when translating a figure along a vector.

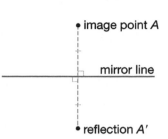

So the vertex A is reflected in the mirror line to give the new vertex A', the vertex B is reflected to give the new vertex B' and the vertex C is reflected to give the new vertex C'.

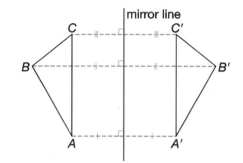

Joining the new vertices A', B' and C' we get $\triangle A'B'C'$, which is the reflection of $\triangle ABC$ in the mirror line.

Constructing lines from each vertex perpendicular to the mirror line involves a lot of work. However, using graph or squared paper makes it much easier as we already have a grid of perpendicular lines.

NOTE:

1 There is no change in the shape or the size of the figure after reflection in normal mirror lines.
2 Any points on the mirror line do not move or change at all during reflection.
   These same points are part of the object and its reflected image, and so the reflection becomes connected with the original object.
   Points that do not move during reflection are called **invariant** (meaning 'never changes') **points**.

Reflection in the *x*-axis and the *y*-axis

We will start by using the horizontal *x*-axis and the vertical *y*-axis on the Cartesian plane as mirror lines.

Example

Reflect the trapezium *ABCD* in the *x*-axis and write down the coordinates of the reflected image.

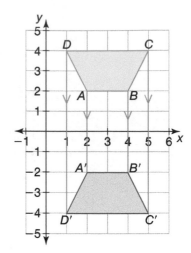

Vertex *A* is 2 units above the *x*-axis (mirror line). So the reflected vertex *A′* will be 2 units on the other side of the *x*-axis, or 2 units below the *x*-axis.
This is also true for vertex *B*: the reflected vertex *B′* will also be 2 units below the *x*-axis.
Vertex *C* is 4 units above the *x*-axis. So the reflected vertex *C′* will be 4 units below the *x*-axis.
This is also true for vertex *D*: the reflected vertex *D′* will also be 4 units below the *x*-axis.

Notice that there is no movement to the left or the right. To find the reflected point we move in a direction that is perpendicular to the mirror line. This means perpendicular to the *x*-axis in this case. Moving perpendicular to the *x*-axis is the same as moving parallel to the *y*-axis, so this is why there is no movement to the left or the right.

The coordinates of the vertices of *ABCD* are: $A(2, 2)$, $B(4, 2)$, $C(5, 4)$ and $D(1, 4)$.
The coordinates of the vertices of the reflection *A′B′C′D′* are: $A'(2, -2)$, $B'(4, -2)$, $C'(5, -4)$ and $D'(1, -4)$.

Can you see the pattern?

> When we reflect any point in the *x*-axis, the *x*-coordinate stays the same and the *y*-coordinate changes sign.

Example

Reflect the triangle *ABC* in the *y*-axis and write down the coordinates of the reflection.

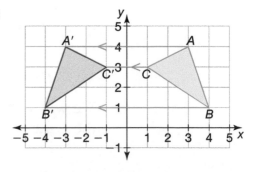

Vertex *A* is 3 units right of the *y*-axis (mirror line). So the reflected vertex *A′* will be 3 units on the other side of the *y*-axis, or 3 units left of the *y*-axis.
Vertex *B* is 4 units right of the *y*-axis. So the reflected vertex *B′* will be 4 units left of the *y*-axis.
Vertex *C* is 1 unit right of the *y*-axis. So the reflected vertex *C′* will be 1 unit left of the *y*-axis.

Notice that there is no movement up or down. To find the reflected point we move in a direction that is perpendicular to the mirror line. This means perpendicular to the *y*-axis in this case. Moving perpendicular to the *y*-axis is the same as moving parallel to the *x*-axis, so this is why there is no movement up or down.

The coordinates of the vertices of *ABC* are: *A*(3, 4), *B*(4, 1) and *C*(1, 3). The coordinates of the vertices of the reflection *A'B'C'* are: *A'*(−3, 4), *B'*(−4, 1) and *C'*(−1, 3).

Can you see the pattern?

> When we reflect any point in the *y*-axis, the *y*-coordinate stays the same and the *x*-coordinate changes sign.

Reflection in any line parallel to the *x*-axis or the *y*-axis

When we reflect a figure in any line that is parallel to either the *x*-axis or the *y*-axis (e.g. the line *y* = 7 or the line *x* = 3), we find a similar pattern occurs with the coordinates of the point and its reflection.

> If the mirror line is parallel to the *x*-axis, the *x*-coordinates of the point and its reflection remain the same.
> If the mirror line is parallel to the *y*-axis, the *y*-coordinates of the point and its reflection remain the same.

The other coordinate will change depending on exactly what the equation of the mirror line is (*y* = *c*, *x* = *b*, etc.).

Example

Reflect the figure *P* in the lines *x* = 4 and *y* = 4.

Reflect each of the vertices as before.

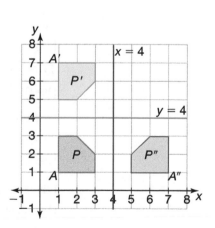

Join the reflected points to draw the reflected image in each case.
Image *P'* is the reflection in the line *y* = 4.
e.g.   Vertex *A*(1, 1) is reflected to the point *A'*(1, 7).

Image *P"* is the reflection in the line *x* = 4.
e.g.   Vertex *A*(1, 1) is reflected to the point *A'*(7, 1).

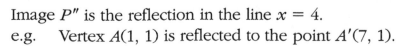

Reflection in the lines $y = x$ and $y = -x$

The line $y = x$ passes through the origin, and any point on it is the same distance from the x-axis as it is from the y-axis.

The line $y = -x$ also passes through the origin, and any point on it is also the same distance from the x-axis as it is from the y-axis.

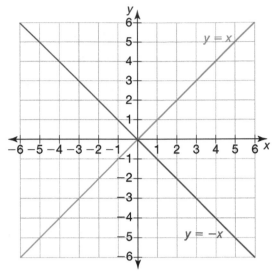

When a point is reflected in the line $y = x$, its coordinates are interchanged. So the x-coordinate becomes the y-coordinate and vice versa.

When a point is reflected in the line $y = -x$, its coordinates are interchanged and they both change signs.

Example

Reflect the figure G in the lines $y = x$ and $y = -x$.

Reflect each of the vertices carefully in the lines $y = x$ and $y = -x$ to give the vertices of the two reflections, G' and G''. (These reflection lines have been shown for only some of the vertices for clarity.)

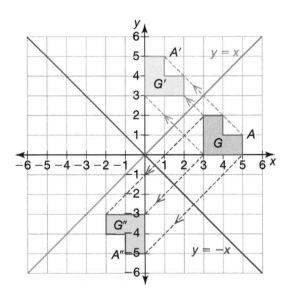

Vertex $A(5, 1)$ is reflected in the line $y = x$ to the point $A'(1, 5)$. The x- and y-coordinates of the point have interchanged.

Vertex $A(5, 1)$ is reflected in the line $y = -x$ to the point $A''(-1, -5)$. The x- and y-coordinates of the point have interchanged and both have changed signs.

Check that the same expected pattern is true for the coordinates of all the vertices and their reflections.

Exercise 3

In questions **1** to **8**, use a Cartesian plane to help you if necessary.

1 Write down the coordinates of the image of each of these points after reflection in the *x*-axis.

a) (3, 6)   b) (−10, −7)   c) (0, 8)   d) (−5, 0)

2 Write down the coordinates of the image of each of these points after reflection in the *y*-axis.

a) (3, 6)   b) (−10, −7)   c) (0, 8)   d) (−5, 0)

3 Write down the coordinates of the image of each of these points after reflection in the line $x = 1$.

a) (4, 7)   b) (−2, −3)   c) (0, 6)   d) (−4, 0)

4 Write down the coordinates of the image of each of these points after reflection in the line $y = 1$.

a) (4, 7)   b) (−2, −3)   c) (0, 6)   d) (−4, 0)

5 Write down the coordinates of the image of each of these points after reflection in the line $y = x$.

a) (9, 8)   b) (−5, −4)   c) (0, −3)   d) (7, 0)

6 Write down the coordinates of the image of each of these points after reflection in the line $y = -x$.

a) (9, 8)   b) (−5, −4)   c) (0, −3)   d) (7, 0)

7 Write down the coordinates of the image of each of these points after reflection in the *x*-axis followed by reflection in the line $y = x$.

a) (7, 2)   b) (−9, −12)   c) (0, −2)   d) (8, 0)

8 Write down the coordinates of the image of each of these points after reflection in the *y*-axis followed by reflection in the line $y = -x$.

a) (7, 2)   b) (−9, −12)   c) (0, −2)   d) (8, 0)

9 Copy each figure onto squared paper and draw its reflection in each of these lines.

   i) the *x*-axis   ii) the *y*-axis   iii) $x = 2$ iv) $y = x$

Write down the coordinates of vertex *B* after each reflection.

a)
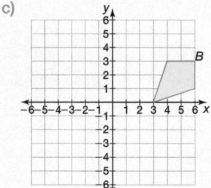

b)
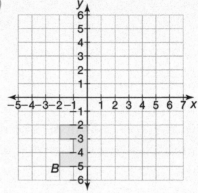

c)
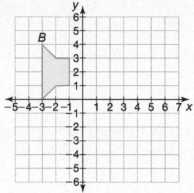

d)
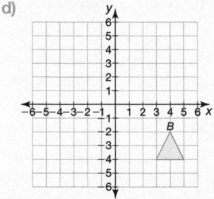

C Rotation

Rotation is circular movement. So the path of the movement draws a circle, or part of a circle, known as an arc.

There are many examples of rotation in everyday life: bicycle wheels, the hands of a clock, a door opening or closing, and so on.

A circle has a centre point, and all circular movement takes place around a fixed point called the **centre of rotation**.

- The centre of rotation can sometimes be a point on the figure that is being rotated.
- The centre of rotation can also be a point some distance away that is not part of the figure.

The amount of rotation is measured in degrees (as in sectors of a circle). This is called the **angle of rotation**.
Rotation can be in a **clockwise** direction (to the 'right') or in an **anticlockwise** direction (to the 'left').

So, for every rotation, we need three pieces of information:

● the centre of rotation
● the amount of the turn (angle of rotation)
● the direction of the turn (clockwise or anticlockwise).

The diagram shows the rotation of a point R through 90° clockwise about the origin.
So $R(-3, 2)$ is mapped onto $R'(2, 3)$ by the rotation.

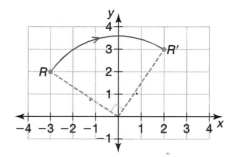

To rotate a whole figure, we simply rotate each vertex in this way, and then join the new vertices to give the image figure. This is the same method as we used to translate or reflect a figure.

Centre of rotation is a point on the figure

Example

Rotate $\triangle PQR$ through 90° anticlockwise about the origin.

Note that the vertex P is at the origin, so the centre of rotation is a point **on** the triangle. P does not move so $P = P'$. We need only rotate the other two vertices, Q and R, to map them onto their image points, Q' and R'.

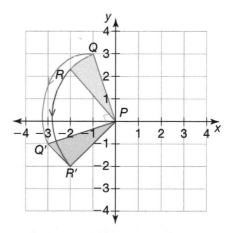

The angle between line segments PQ and PQ' is 90°. The angle between line segments PR and PR' is 90°.

Each of these 90° angles must be drawn accurately using either a protractor or the construction methods you learned in Coursebook 1 Unit 10.

The length of PQ is the same as the length of $P'Q'$, and the lengths of PR and $P'R'$ are also equal.

Centre of rotation is a point outside the figure

Rotate $\triangle PQR$ through 180° about the origin.

This time the centre of rotation is **not** a point on the triangle.

Join each vertex of the triangle to the centre of rotation (origin) with a line (shown as the different coloured lines).

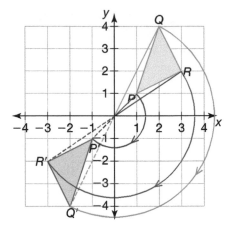

Construct an angle of 180° on each of these lines and extend the lines an equal distance to find the image point of each vertex.

Join the image points to draw the image $\triangle P'Q'R'$.

NOTE: These methods are also used when the centre of rotation is at any point, and not simply at the origin.

Steps for constructing the rotation of a figure about a point

1 Join the centre of rotation to each vertex of the figure with a straight line.
2 At the centre of rotation, construct on each of these lines the angle of rotation given and in the direction given.
3 Use a pair of compasses to measure each of the lines and mark the corresponding constructed angle with the same length to find the image points.
4 Label these images of the vertices.
5 Join the image points to give the rotated figure.

 Exercise 4

1 Use a construction on squared paper to work out the coordinates of the image for each of these rotations.

a) The point (9, −8) is rotated 90° clockwise about the origin.
b) The point (−3, 5) is rotated 90° anticlockwise about the origin.
c) The point (5, −7) is rotated 180° about the origin.
d) The point (2, 1) is rotated 90° clockwise about the origin.
e) The point (0, 3) is rotated 180° about the origin.
f) The point (−1, 1) is rotated 270° clockwise about the origin.

2 Use a construction on squared paper to work out the coordinates of the image for each of these rotations.

a) The point $(-2, -3)$ is rotated 90° clockwise about the point $(2, 1)$.
b) The point $(0, -2)$ is rotated 90° anticlockwise about the point $(-1, 2)$.
c) The point $(5, 3)$ is rotated 180° about the point $(1, 3)$.

3 Construct the image of each of these line segments after the rotation given and write down the coordinates of A' and B'.

a) 90° anticlockwise about the origin

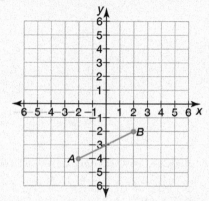

b) 180° about the origin

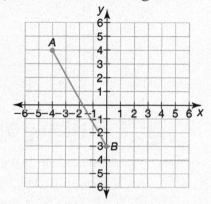

c) 90° clockwise about the point $(1, 2)$

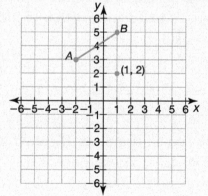

d) 180° about the point $(2, -1)$

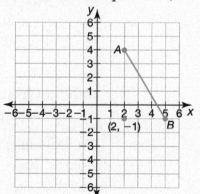

4 Construct the image of each of these figures after the rotation given and write down the coordinates of B'.

a) 90° anticlockwise about the origin

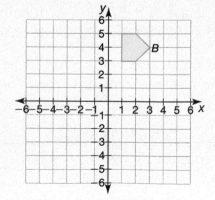

b) 180° about the origin

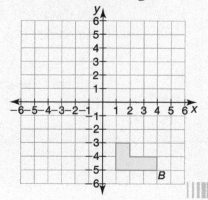

c) 90° clockwise about the point (2, 0)

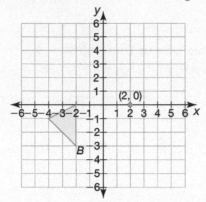

d) 180° about the point (1, 2)

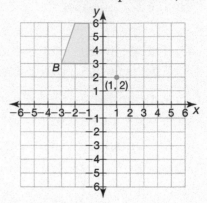

D Enlargement

So far we have learned about translation, reflection and rotation. These transformations do not change the size or the shape of the figures that are transformed. We call this kind of transformation **isometric**. This name means 'equal measurement' and comes from two Greek words: *iso* which means 'equal', and *metreo*, which we have already learned means 'to measure'.

Enlargement means 'to make bigger' and so the size of the image is different from the size of the object after this transformation. This means that enlargement is **not** an isometric transformation.

Enlargement is something that we see all the time. Big advertising boards often show very big photographs. These photos did not start out this size. The small negative (or digital image) is enlarged until it is the size needed. An overhead projector also enlarges what is written on the transparency so that it is easy to read, even at the back of the room.

Enlargement does not change the shape of the figure. All the angles stay the same. It is only the length of the sides in the figure that change.

It is important that we know how much bigger the image will be. We call this the **scale factor**. The scale factor tells us how many times bigger than the object the image will be. The scale factor is defined as:

$$\text{scale factor} = \frac{\text{length of image}}{\text{corresponding length of object}}$$

So, if the figure is a triangle, then we calculate the scale factor using the lengths of the corresponding sides in the object and image.

Even though the word 'enlargement' means to make bigger, in maths we use the same word even if we are making a figure smaller. When the scale factor is a fraction less than 1 there is a **reduction** in size because the image length will be smaller than the object length.

Constructing an enlargement

To construct an enlargement of any figure, we use a special point called the **centre of enlargement**. This can be any convenient point, and it can be outside or inside the object figure.

Example

Enlarge $\triangle ABC$ three times (scale factor 3) with the origin as the centre of enlargement.

Join each of the vertices of $\triangle ABC$ to the centre of enlargement, O, with straight lines and extend these straight lines some distance beyond the vertices.

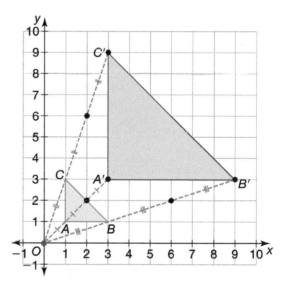

Use a pair of compasses to measure OA and mark A' so that $OA' = 3 \times OA$.
Measure OB and mark B' so that $OB' = 3 \times OB$.
Measure OC and mark C' so that $OC' = 3 \times OC$.
Join the points A', B' and C' to draw the enlarged $\triangle A'B'C'$.
Check that each side of the image $A'B'C'$ is 3 times longer than the corresponding side of the object ABC.
$A'B' = 3 \times AB$, $A'C' = 3 \times AC$ and $C'B' = 3 \times CB$

The next example has a centre of enlargement that is not the origin, and is also inside the figure.

Example

Enlarge the parallelogram $ABCD$ by a scale factor of 2 with the point $(5, 5)$ as the centre of enlargement.

We can use the same method as we used in the previous example, except the centre of enlargement is now inside the object.

Join each of the vertices of ABCD to the centre of enlargement, X, and extend these lines some distance beyond the vertices.
Use a pair of compasses to measure XA and mark A' so that $XA' = 2 \times XA$.
Measure XB and mark B' so that $XB' = 2 \times XB$.
Measure XC and mark C' so that $XC' = 2 \times XC$.
Measure XD and mark D' so that $XD' = 2 \times XD$.

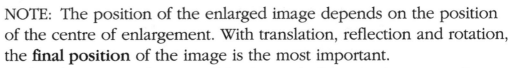

Join the points A', B', C' and D' to draw the enlarged parallelogram $A'B'C'D'$.
Check that each side of the image is twice as long of the corresponding side of the object.

NOTE: The position of the enlarged image depends on the position of the centre of enlargement. With translation, reflection and rotation, the **final position** of the image is the most important.
With enlargement, the **size** of the image is the most important. Its position will change depending on which point is chosen as the centre of enlargement.

Activity

Enlarge the parallelogram in the example above by a scale factor of 2 but use the point (1, 2) as the centre of enlargement. You will find the image is the same size as the image in the example but it will be in a different place.

Example

The square ABCD is enlarged by a scale factor of $\frac{1}{3}$ with the point (0, 4) as the centre of enlargement. Find the coordinates of the image square $A'B'C'D'$. (As the scale factor is less than 1, the image will be **smaller** than the object.)

The method is similar to the one we have used so far, with only some small differences.
Join each of the vertices of ABCD to the centre of enlargement, X.
Use a pair of compasses to measure XA and mark A' so that $XA' = \frac{1}{3} \times XA$.

Measure XB and mark B' so that $XB' = \frac{1}{3} \times XB$.

Measure XC and mark C' so that $XC' = \frac{1}{3} \times XC$.

Measure XD and mark D' so that $XD' = \frac{1}{3} \times XD$.

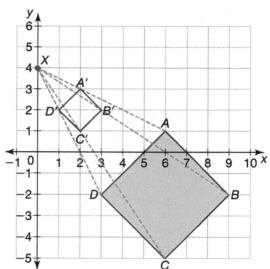

Join the points A', B', C' and D' to draw the 'enlarged' square $A'B'C'D'$.

Check that each side of the image is $\frac{1}{3}$ as long as the corresponding side of the object.

The coordinates are: $A'(2, 3)$, $B'(3, 2)$, $C'(2, 1)$ and $D'(1, 2)$.

Finding the centre of enlargement and the scale factor

If we are given both the object and the image, it is possible to find the centre of enlargement and the scale factor.

1 Join the corresponding vertices of the object and the image figures with straight lines.
2 Extend these lines until they all meet at one point. This point is the centre of enlargement.
3 Measure the length from the centre of enlargement to corresponding vertices on the object and image figures.
4 Calculate the scale factor $= \dfrac{\text{image length}}{\text{object length}}$

Example

You are given the square $ABCD$ and its image $A'B'C'D'$.
Find the centre of enlargement and the scale factor.

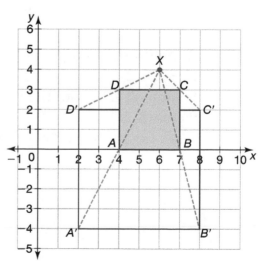

Join the corresponding vertices of the object and image.
Extend these lines until they meet at the point X.
This point $X(6, 4)$ is the centre of enlargement.

Measure the lengths of XA and XA'.

Calculate the scale factor $= \dfrac{XA'}{XA} = 2$

Exercise 5

1   Copy each of these figures onto squared paper and then enlarge it using the centre of enlargement and scale factor given.

   a) Centre (0, 0)
   Scale factor 2

   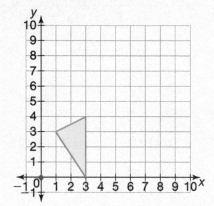

   b) Centre (5, 5)
   Scale factor 3

   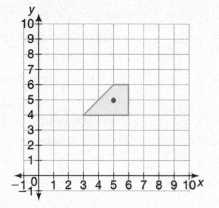

   c) Centre (0, 0)
   Scale factor $\frac{1}{3}$

   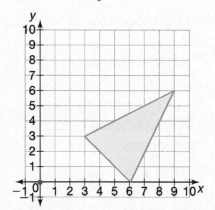

   d) Centre (5, 5)
   Scale factor $\frac{1}{4}$

   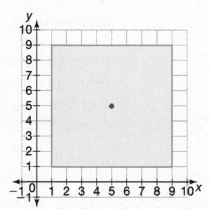

2   Draw on squared paper the triangle ABC with vertices $A(-4, 0)$, $B(0, 2)$ and $C(-2, -2)$.
   Enlarge $\triangle ABC$ with the point A as the centre of enlargement and a scale factor of 2.
   Write down the vertices of the image $\triangle A'B'C'$.

3 The diagram shows the parallelogram *ABCD*.

Copy this parallelogram onto squared paper.
Work out the coordinates of the vertex *C* after
each of these enlargements.

a) Centre (0, 0), scale factor 2
b) Centre (0, 0), scale factor 3
c) Centre (0, 2), scale factor 2
d) Centre *A*(1, 1), scale factor 2
e) Centre *B*(4, 1), scale factor 3
f) Centre *D*(2, 3), scale factor 2

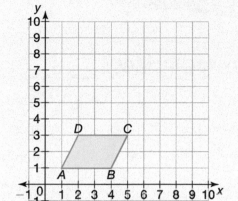

4 For each of these diagrams, find the scale factor and the coordinates of the
centre of enlargement.

a)
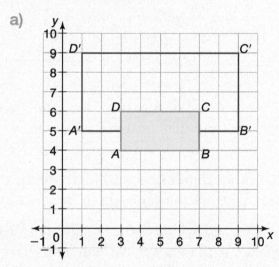

b)
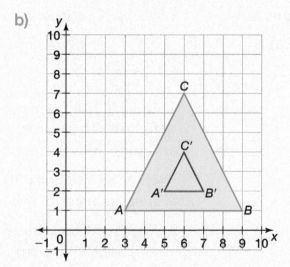

c)
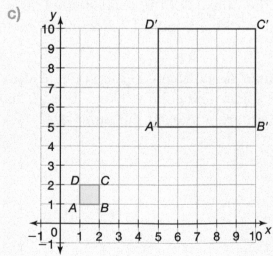

Geometry in three dimensions

In Unit 1 of Coursebook 2 we learned to measure plane geometric figures. These are the figures that are 'flat' and are drawn in two **dimensions**.

Now we are going to add another dimension and learn about the measurement of solid figures that are drawn in three dimensions.

We can already measure length and width and calculate perimeters and areas. For shapes that have a 'thickness', we will learn to calculate the total area on the outside of the solid shape (called the **surface area**) and the **volume** (the amount of space inside the solid shape).

The simplest solid three-dimensional shape is the **cube**. You may have seen a cube of ice or sugar. The cube is like a 'box'. It has flat sides, each the shape of a **square** and each the same size.

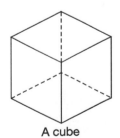

A cube

You can see from the drawing that a cube has **six sides** and if we 'unfold' these sides so that they lie flat on the paper, we would see this:

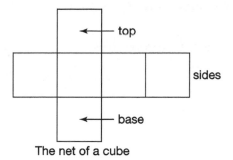

The net of a cube

This two-dimensional, flat drawing of all the sides of a solid three-dimensional shape is called the net of the shape.

Our paper is flat and two-dimensional, so it is not easy to draw a clear picture on paper of a three-dimensional solid shape. Sometimes it is helpful to colour in some of the sides with **shading** as if light is shining on the shape from one direction, as an artist does to make pictures look 'lifelike'.

In maths we use a special paper with a triangular grid pattern to help us draw the angles correctly so that we can 'see' the three-dimensional shapes on paper.

Sometimes the paper has dots to show the triangular pattern.

Sometimes the paper has a full triangular grid.

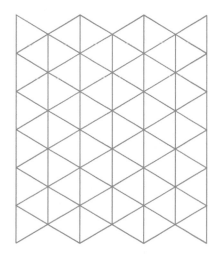

Paper with this kind of triangular grid pattern is called isometric paper.

Use these grids to help you draw a cube.

- In a solid shape, we call every flat surface a **face**.
- The line where two faces meet is called an **edge**.
- The point at a 'corner' where edges meet is called a **vertex**.

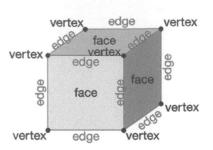

 Exercise 1

1 Here is one way to draw a net of a cube.
There are 12 different ways to draw the net of a cube.
Use squared paper to draw all 12 different nets.
Cut out each of the nets you draw and fold them to make a cube.
This will help you to check that it is correct, and that it is different from the others you have drawn.

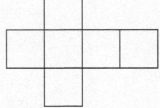

2 Use squared paper to draw an accurate net of each of these three-dimensional shapes.

a)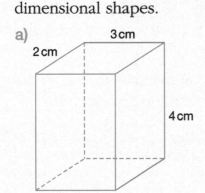

b)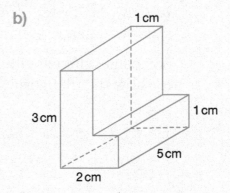

c)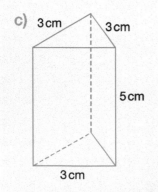

d)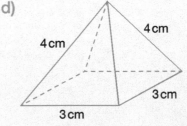

3 Start with four cubes each 1 cm by 1 cm by 1 cm. Try to make as many different shapes as you can by joining the cubes face-to-face. There are eight different shapes altogether.
Use isometric paper to draw a three-dimensional drawing of each of the shapes.

A Cuboids

A cube is a solid three-dimensional shape with six identical flat faces, each a square of the same size.

A cuboid is the name for any three-dimensional shape that has six flat faces that are **rectangular**.

So a cube is a special example of a cuboid.
To measure a cuboid, we need to measure in three dimensions (or directions).
We call these the length, width and height of the cuboid.

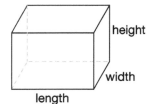

Volume of a cuboid

To calculate the volume of a cuboid, we simply multiply these three measurements together.

> Volume of a cuboid = length × width × height
> *or* $V = lwh$
> Volume of a cube = length × length × length
> *or* $V = l^3$

From this formula, we can see that the units of volume will be (a unit of length)^3.

The most common unit of volume is **cm³** which is read as 'cubic **centimetre**'.

Smaller volumes can be measured in **mm³** which is read as 'cubic **millimetre**'.

Larger volumes can be measured in **m³** which is read as 'cubic **metre**'.

Surface area of a cuboid

The surface area of a cuboid is the total of the area of each of its rectangular faces. The simplest way to calculate the surface area is first to draw the net of the cuboid. Then you can calculate the area of each of the rectangular faces, and add them together.

Example Calculate the surface area and the
volume of this cuboid.

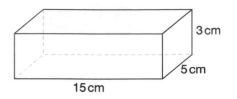

First draw the net of the cuboid.

15 cm

5 cm | 15 cm | 5 cm

3 cm | 15 cm | 3 cm

3 cm
5 cm | 5 cm | 15 cm | 5 cm | 5 cm
3 cm

3 cm | 15 cm | 3 cm

Area of all the rectangular faces
$= (15 \times 5) + (15 \times 3) + (15 \times 5) + (15 \times 3) + (5 \times 3) + (5 \times 3)$
$= 2(15 \times 5) + 2(15 \times 3) + 2(5 \times 3)$
$= 150 + 90 + 30$
$= 270$
The surface area of the cuboid is $270 \, cm^2$.

The volume of the cuboid $=$ length $\times$ width $\times$ height
$= 15 \times 5 \times 3$
$= 225$
The volume of the cuboid is $225 \, cm^3$.

Exercise 2

1 Copy and complete this table by calculating the missing values for each of
the cuboids.

	Length	Width	Height	Surface area	Volume
a)	3 cm	3 cm	3 cm		
b)	5 m		2 m		30 m³
c)	30 mm	15 mm		1980 mm²	
d)	6 cm	3.6 cm	2.4 cm		
e)		3.2 m	4.8 m		96.8 m³
f)	3.2 cm	2.4 cm	10 cm		
g)		7.5 cm	12 cm	1155 cm²	

B Prisms

A **prism** is a three-dimensional shape. It has two faces exactly the same shape, parallel and opposite to each other. If we cut 'slices' parallel to these faces, all the 'slices' will have the same shape and size. We call this 'slice' a **cross-section** of the prism.

All the shapes below are examples of prisms. You can see that a cuboid is also a prism. It is a very special prism and that is why we studied it separately.

Can you draw other shapes that are prisms?

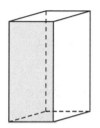 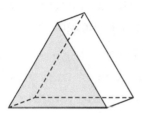  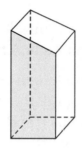 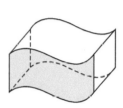

Volume of a prism

> **Volume of prism** = area of the cross-section × length

Cylinders

A **cylinder** is a special prism with a cross-section that is a **circle**. The formula for the volume of a cylinder is the same as for a prism. The cross-section of a cylinder is a circle, and the area of a circle is πr^2. We normally talk about the height of a cylinder rather than the length.

> **Volume of cylinder** = area of the cross-section × height
> $= \pi r^2 h$

To work out the surface area, draw the net of the cylinder. The top and bottom of a cylinder are circles.

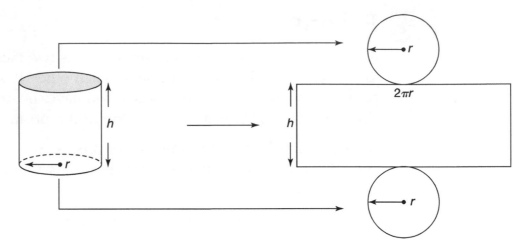

If we 'unroll' the curved surface (like a carpet), it will form a **rectangle**. The rectangle has the same height, h, as the cylinder.

The length of the rectangle must be just long enough to 'wrap around' the circle. The top and bottom of the cylinder each have radius r and circumference of $2\pi r$. So the length of the rectangle is also $2\pi r$.

Area of the top $= \pi r^2$
Area of the base $= \pi r^2$
So the area of the top and base $= 2\pi r^2$

Area of the rectangle $=$ length $\times$ width
$$= 2\pi r \times h$$
So area of curved surface $= 2\pi rh$

For a **cylinder** with radius, r, and height, h,
surface area $=$ area of top and base
$+$ area of curved surface
$$= 2\pi r^2 + 2\pi rh$$
$$= 2\pi r(r + h)$$

Examples — Find the volume of each of these prisms.

a)

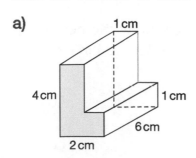

First we need to find the area of the shaded cross-section of the prism:

Area $= (4 \times 1) + (1 \times 1)$
$\quad\;\; = 4 + 1 = 5$

The area of the cross-section is $5\,cm^2$.

Volume of prism = area of cross-section $\times$ length
$\qquad\qquad\qquad = 5 \times 6 = 30$

The volume of the prism is $30\,cm^3$.

b)

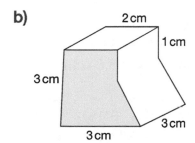

Area of shaded cross-section = area of rectangle + area of triangle
$$= (3 \times 2) + \left(\tfrac{1}{2} \times 1 \times 2\right)$$
$$= 6 + 1 = 7$$

The area of the cross-section is $7\,cm^2$.

Volume of prism = area of cross-section $\times$ length
$\qquad\qquad\qquad = 7 \times 3 = 21$

The volume of the prism is $21\,cm^3$.

Examples

Find the volume and surface area of these cylinders (use $\pi = 3.14$).

a)

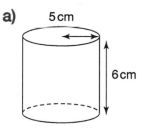

Volume $= \pi r^2 h$
$\qquad\quad = \pi \times 5^2 \times 6$
$\qquad\quad = 3.14 \times 25 \times 6 = 471$

The volume of the cylinder is $471\,cm^3$.

Surface area $= 2\pi r(r + h)$
$\qquad\qquad\quad = 2 \times 3.14 \times 5(5 + 6)$
$\qquad\qquad\quad = 31.4 \times 11 = 345.4$

So the surface area of the cylinder is $345.4\,cm^2$.

b)

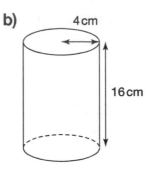

Volume $= \pi r^2 h$
$= \pi \times 4^2 \times 16$
$= 3.14 \times 16 \times 16 = 803.84$
The volume of the cylinder is $803.84 \, \text{cm}^3$.

Surface area $= 2\pi r(r + h)$
$= 2 \times 3.14 \times 4(4 + 16)$
$= 25.12 \times 20 = 502.4$
The surface area of the cylinder is $502.4 \, \text{cm}^2$.

Exercise 3

1 Calculate the volume of each of these prisms correct to 1 decimal place (use $\pi = 3.14$).

a)

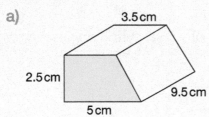

b)

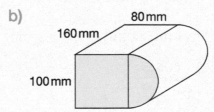

c)

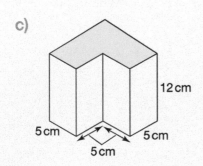

d)

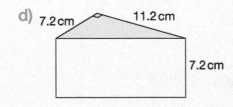

2 Copy and complete this table by calculating the missing values for each of the cylinders correct to 1 decimal place (use $\pi = 3.14$).

	Radius	Diameter	Height	Area of curved surface	Total surface area	Volume
a)	3 cm		4 cm			
b)	4 cm					201.0 cm³
c)		6 m		282.6 m²		
d)			50 mm			62 800 mm³
e)	2.7 cm			135.6 cm²		
f)		10.6 cm				1058.4 cm³
g)		6.5 m	1.2 m			

 3 A bucket in the shape of a cylinder with radius 15 cm and height 40 cm is full of water.
The water is poured into a cylindrical bowl with a diameter of 40 cm.
Calculate the height of the water in this bowl (use $\pi = 3.14$).

C Pyramids

A **pyramid** is a solid three-dimensional figure with a base that is a polygon and the **slant faces** are all triangles that meet at one vertex point above the base.

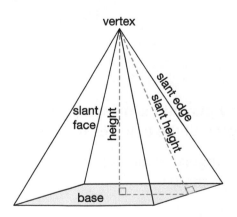

- Each vertex point on the base is joined to the common vertex above the base by a line called the slant edge.
- The perpendicular distance from the edge of the base to the common vertex point is called the slant height.
- The perpendicular distance from the common vertex point to the base is called the **height** (or perpendicular height).
- If the common vertex point is directly above the centre point of the base then the pyramid is called a right pyramid.
- If the base of a right pyramid is a regular polygon then all slant faces of this pyramid are equal.

The name of a pyramid comes from the polygon that it has for a base. So a pyramid with a rectangle for a base will be called a rectangular pyramid, and so on.

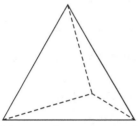

Triangular pyramid
or **tetrahedron**

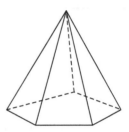

Pentagonal pyramid

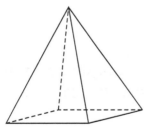

Rectangular pyramid
or square pyramid

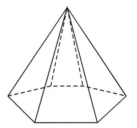

Hexagonal pyramid

The surface area of a pyramid

The **surface area of a pyramid** is made up of the area of the base shape, plus the area of all the triangles that make the slant faces.

- The number of triangular slant faces will be the same as the number of sides on the base.

- The base of each triangle will be the same length as one side of the base polygon.
- The height of each triangle will be the 'slant height' of the pyramid.
- If the pyramid is a right pyramid, then all the slant faces will be isosceles triangles.

For example, the surface area of a right square pyramid will be the area of the square base plus the area of four isosceles triangles.

Example

A right rectangular pyramid has a base with sides of 6 cm by 8 cm. The slant heights of the triangular faces are 10 cm and 10.3 cm. Calculate the total surface area of the pyramid.

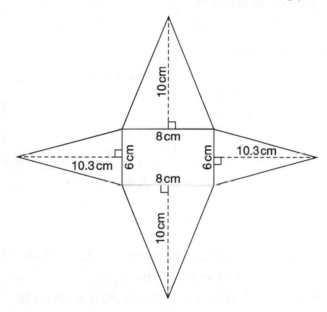

The **net** of this rectangular pyramid will be a rectangle and two pairs of equal isosceles triangles.

The total surface area of the pyramid
= area of base + area of four triangles
$$= (6 \times 8) + 2 \times \left(\tfrac{1}{2} \times 8 \times 10\right) + 2 \times \left(\tfrac{1}{2} \times 6 \times 10.3\right)$$
$$= 48 + 80 + 61.8 = 189.8$$

The total surface area of the pyramid is 189.8 cm^2.

The volume of a pyramid

Activity

Your teacher will be able to show you
how to cut out some cardboard as shown in
this diagram.
You can choose your own length for *a* and
for the other sides of the isosceles triangles in the
net.
Fold along the dotted lines and put some glue on
the grey strip.
This will make a tetrahedron with an open base.
Measure its height, *h*.

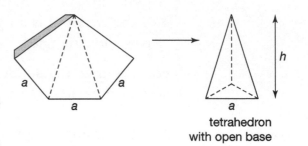

tetrahedron
with open base

Now cut out another piece of cardboard as
shown in the second diagram.
Use the same length for *a* as you did for the
tetrahedron and use the same length for *h* that
you measured on you tetrahedron.
Fold along the dotted lines and put some glue
on the grey strips.
This will make a triangular prism with one end
open and the other end closed.

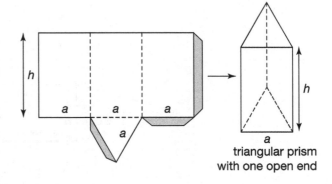

triangular prism
with one open end

The triangular base of the tetrahedron should be the same size as the base of the prism.
Fill the tetrahedron with sand and pour the sand into the prism.
Repeat this until the prism is full of sand. How many times did you need to fill the tetrahedron?

Through experiments like this, we can show that the volume of any
pyramid is equal to one-third the volume of a prism that has the same
size and shaped base and the same perpendicular height as the
pyramid.

We already know how to calculate the volume of prism:
volume = area of the cross-section (or base)
× length (or perpendicular height)

Volume of a pyramid $= \frac{1}{3}Ab$
where *A* is the area of the base and *b* is the
perpendicular height of the pyramid

Examples

a) Find the volume of a right rectangular pyramid if its base is 5 cm by 3 cm and its height is 8 cm.

Volume of a pyramid $= \frac{1}{3}Ah$

$$= \frac{1}{3}(5 \times 3) \times 8$$
$$= 5 \times 8 = 40$$

The volume of the pyramid is $40 \, cm^3$.

b) Find the height of a right tetrahedron if it has a volume of $147 \, cm^3$ and a base area of $63 \, cm^2$.

Volume of tetrahedron $= 147$

$$\frac{1}{3}Ah = 147$$
$$h = 147 \div A \times 3$$
$$= \frac{147 \times 3}{63}$$
$$= 7$$

The height of the tetrahedron is 7 cm.

c) A square pyramid of height 12 cm has a volume of $784 \, cm^3$.
Calculate:
  i) the area of the base
  ii) the length of each side of the base.

i) Volume of pyramid $= 784$

$$\frac{1}{3}Ah = 784$$
$$\frac{1}{3} \times A \times 12 = 784$$
$$A = \frac{784 \times 3}{12}$$
$$= 196$$

The area of the base is $196 \, cm^2$.

ii) The base is a square.

So $l^2 = 196$
$$l = \sqrt{196} = 14$$

The length of the side is 14 cm.

Cones

When we first described a circle in Coursebook 2, we said it is almost like a polygon with so many sides that we cannot count them. The sides of the polygon become so small that the shape is now a circle.

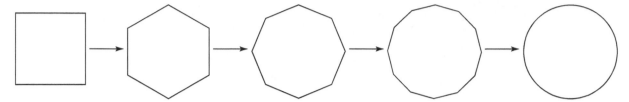

If each of these polygons is the base of a pyramid, then we can think of a **cone** as a pyramid with so many faces that we cannot count them.

So a cone is really a pyramid whose base is a **circle** rather than a polygon.

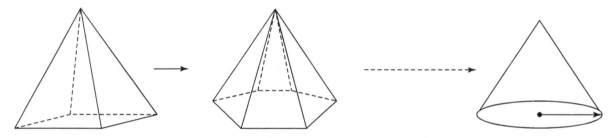

We have seen a cone-shaped object before: a paper cup, an ice-cream cone. If we cut open a regular cone shape from the wide edge to the 'point' or vertex, we find that the **curved surface** of the cone opens out to become a sector (or slice) of a circle.

The radius of this sector is the same as the slant height of the cone.

So we really have **two different circles** (or parts of them) in a cone:

- The smaller circle is the base. The radius of this circle is usually called r.
- The curved surface is part of a bigger circle. The radius of this circle is usually called l as it is the same as the slant height of the cone.

If the circular base of the cone has a radius r, then the circumference of the circular base will be $2\pi r$. This length is the same as the **arc length** of the sector of the bigger circle when we open up the curved surface of the cone.

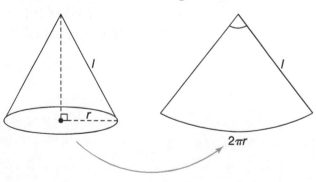

The surface area of a cone

The surface area of a cone is made up of two parts:

- the area of the smaller circle on the base
- the area of the curved surface (a sector of the bigger circle).

We learned in Unit 6 that the ratio of the area of a sector to the area of the whole circle is equal to the ratio of the angle of the sector to the angle in a full circle (360°). Also, the ratio of the arc length of a sector to the full circumference of the circle is equal to the ratio of the angle of the sector to the angle in a full circle (360°).

These two facts mean that the following equation is also true about a sector and its circle:

$$\frac{\text{area of sector}}{\text{area of circle}} = \frac{\text{arc length of sector}}{\text{circumference of circle}}$$

We can use this equation to find the area of the sector that is the curved surface of the cone.

$$\frac{\text{area of sector}}{\pi l^2} = \frac{2\pi r}{2\pi l}$$

$$\text{area of sector} = \frac{2\pi r}{2\pi l} \times \pi l^2 = \pi r l$$

The area of this sector is the same as the area of the curved surface of the cone.

Area of the curved surface of a cone $= \pi r l$
where r is the radius of the base and l is the slant height.

Total surface area of a cone = area of curved surface
+ area of base circle
$= \pi r l + \pi r^2 = \pi r(l + r)$

Example

Find the total surface area of a solid cone of base radius 9 cm and slant height 26 cm. $\left(\text{Use } \pi = \frac{22}{7}.\right)$

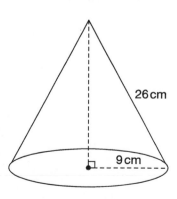

Total surface area = area of curved surface
+ area of base
$= \pi r l + \pi r^2 = \pi r(l + r)$
$= \frac{22}{7} \times 9 \times (26 + 9)$
$= \frac{22}{7} \times 9 \times 35 = 990$

So total surface area is 990 cm².

Examples

a) Find the slant height of a cone which has a curved surface area of 23.47 cm² and base radius 1.9 cm. $\left(\text{Use } \pi = \frac{22}{7}.\right)$

Curved surface area = 23.47
$$\pi r l = 23.47$$
$$l = 23.47 \times \frac{7}{22} \times \frac{1}{1.9} = 3.9$$
The slant height is 3.9 cm.

b) A cone has a curved surface area of 42.5 cm² and a slant height of 4.1 cm.
Calculate the area of the base of this cone. $\left(\text{Use } \pi = \frac{22}{7}.\right)$

Curved surface area = 42.5
$$\pi r l = 42.5$$
$$r = 42.5 \times \frac{7}{22} \times \frac{1}{4.1}$$
$$= 3.3$$
The radius of the base is 3.3 cm.
Area of base $= \pi r^2$
$$= \frac{22}{7} \times 3.3 \times 3.3 = 34.2$$
The area of the base of the cone is 34.2 cm².

c) A cone is made out of a quadrant $\left(\frac{1}{4} \text{ of a circle}\right)$ with a radius of 36 cm. Calculate the radius of the base of this cone.

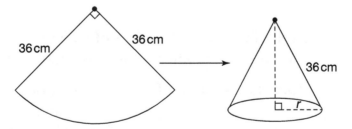

Length of the arc $= \frac{1}{4} \times 2\pi(36)$

Circumference of cone base = length of arc
$$2\pi r = \frac{1}{4} \times 2\pi(36)$$
$$r = \frac{2\pi \times 36}{4 \times 2\pi}$$
$$= 9$$
The radius of the base of the cone is 9 cm.

The volume of a cone

We already know the formula for calculating the volume of a pyramid:

Volume of pyramid $= \frac{1}{3} \times$ area of base $\times$ perpendicular height

We can use this to find the formula for calculating the volume of a cone where the base is a circle, with area $= \pi r^2$.

> **Volume of a cone** $= \frac{1}{3}\pi r^2 h$
> where r is the radius of the base and h is the perpendicular height of the cone.

So the volume of a cone is $\frac{1}{3} \times$ the volume of a **cylinder** with the same radius and height.

Examples

a) Find the volume of a solid cone with a base radius of 3.5 cm and a perpendicular height of 18 cm. $\left(\text{Use } \pi = \frac{22}{7}.\right)$

Volume of cone $= \frac{1}{3}\pi r^2 h$

$= \frac{1}{3} \times \frac{22}{7} \times \frac{7}{2} \times \frac{7}{2} \times 18 = 231$

So volume of cone is 231 cm³.

b) Find the height of a solid cone with a base radius of 11 cm and a volume of 1573π cm³.

Volume of cone $= 1573\pi$ cm³

$\frac{1}{3}\pi r^2 h = 1573\pi$

$h = 1573\pi \times 3 \times \frac{1}{\pi} \times \frac{1}{11} \times \frac{1}{11} = 39$

The height of the cone is 39 cm.

c) Find the radius of the base of a solid cone with a height of 4 cm and a volume of 150.72 cm³. (Use $\pi = 3.14$.)

Volume of cone $= 150.72$ cm³

$\frac{1}{3}\pi r^2 h = 150.72$

$r^2 = 150.72 \times 3 \times \frac{1}{4} \times \frac{1}{3.14} = 36$

$r = \sqrt{36} = 6$

The radius of the base of the cone is 6 cm.

Exercise 4

1 For each of these solid figures, draw the net and calculate its total surface area. Use π = 3.14.

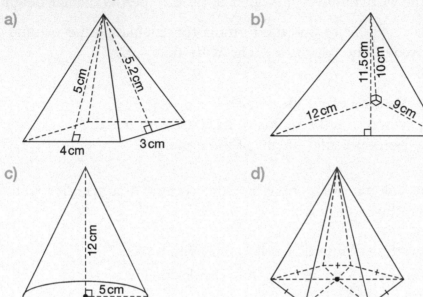

a)

5 cm   5.2 cm

4 cm   3 cm

b)

11.5 cm   10 cm

12 cm   9 cm

c)

12 cm

5 cm

d)

8 cm

Slant height = 14 cm
Perpendicular distance from centre of base to one of its sides = 2.3 cm

2 Calculate the missing values for each of these pyramids and complete a copy of the table.

	Height	Base area	Volume
a)	8 cm		168 cm³
b)		36 cm²	72 cm³
c)	12 cm	70 cm²	

3 Calculate the missing values for each of these square pyramids and complete a copy of the table.

	Length of side in base	Height	Base area	Volume
a)	13 cm	12 cm		
b)		9 cm		75 cm³
c)			49 cm²	98 cm³

4 A pyramid with a triangular base has a volume of 50 cm³. The base and height of the triangle are 5 cm and 8 cm.
Calculate the height of the pyramid.

5 Calculate the missing values for each of these cones, correct to 1 d.p., and complete a copy of the table. Use $\pi = 3.14$.

	Base radius (cm)	Base diameter (cm)	Height (cm)	Slant height (cm)	Base area (cm²)	Volume (cm³)	Curved surface area (cm²)	Total surface area (cm²)
a)	7		24	25				
b)					78.5	314		282.6
c)			9	15	452.2			
d)				26		6028.8	1959.4	
e)		24.8				1496.7		1086.3
f)	6.4		6					369.8
g)			17	20		499.8		
h)			16.8	14			246.2	

6 The semicircle shown in the diagram is folded to form a cone with an open base.

a) Calculate the radius of the base of this cone.
b) Calculate the height of the cone correct to 1 d.p.
c) Calculate the volume of the cone correct to 1 d.p.
(Use $\pi = 3.14$.)

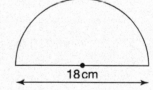

18 cm

7 The diagram shows two solid figures.
Figure A is made up of a square prism joined to a pyramid with the same base.
Figure B is a cylinder joined to a cone with the same base.
Both figures are the same height.
Calculate the volume of each figure correct to 1 d.p. to decide which one is bigger.
Use $\pi = 3.14$.

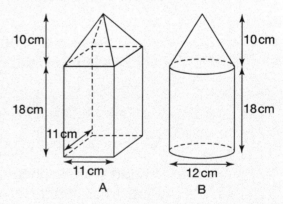

D Spheres

You will have seen many examples of a **sphere**. If you play a game such as football, basket ball or tennis, the ball you use is in the shape of a sphere. (A rugby ball is not a sphere.)

The surface area of a sphere

Many hundreds of years ago, a mathematician in ancient Greece did some experiments and showed that the surface area of a sphere is the same as the curved surface area of a cylinder with the same diameter as the sphere and a height also equal to this diameter.

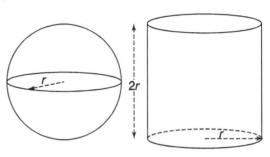

To see this for yourself, carry out this simple experiment.

- Find half a sphere (we call this a **hemisphere**) and a cylinder that have the same radius. (A hemisphere will stay firmly on the table as it has a flat surface.)
- Wind some string or rope carefully around the hemisphere.
- Then take this same length of rope and wind it carefully around the curved surface of the cylinder.
- You will find that the rope will cover the cylinder up to a height that is the same as the radius of the sphere and the cylinder.

The curved surface area of a cylinder $= 2\pi rh$ and, if $h = 2r$, this surface area is equal to the surface area of a sphere with the same radius. So the surface area of a sphere $= 2\pi r(2r) = 4\pi r^2$

> **Surface area of a sphere $= 4\pi r^2$,**
> where r is the radius.

The volume of a sphere

We can carry out another experiment to find the volume of a sphere.

- Start with a sphere that has a radius r, and a hollow cylinder also with a radius r and a height of $2r$.

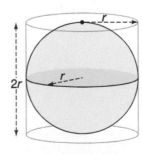
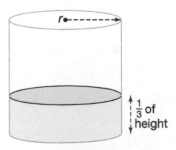

- Place the sphere inside the cylinder and then add water until the cylinder is full.
- Carefully remove the sphere and measure the height of the water left in the cylinder.
- You will find that the water fills exactly $\frac{1}{3}$ of the cylinder.

This means that the volume of the sphere is the same as $\frac{2}{3}$ of the volume of the cylinder.

Volume of sphere $= \frac{2}{3} \times$ volume of cylinder
$$= \frac{2}{3} \times \pi r^2 h = \frac{2}{3} \times \pi r^2 (2r)$$
$$= \frac{4}{3}\pi r^3$$

> **Volume of a sphere** $= \frac{4}{3}\pi r^3$,
> where r is the radius.

Examples

a) A sphere has a radius of 2.5 cm. Use $\pi = 3.14$ to calculate the following correct to 1 d.p.

   i) Calculate the surface area.

   Surface area of a sphere $= 4\pi r^2$
   $$= 4 \times 3.14 \times 2.5 \times 2.5 = 78.5$$
   The surface area of the sphere is 78.5 cm².

   ii) Calculate the volume.

   Volume of a sphere $= \frac{4}{3}\pi r^3$

   $$= \frac{4}{3} \times 3.14 \times 2.5 \times 2.5 \times 2.5$$
   $$= 65.4$$
   The volume of the sphere is 65.4 cm³.

b) A solid hemisphere has a volume of 486π cm³.
   Use $\pi = 3.14$ to calculate the following.
   i) Calculate the radius.

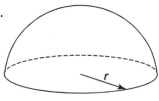

   Volume of hemisphere $= \frac{1}{2} \times \frac{4}{3}\pi r^3$

   $$\frac{1}{2} \times \frac{4}{3}\pi r^3 = 486\pi$$
   $$r^3 = 486 \times 2 \times \frac{3}{4}$$
   $$= 729$$
   $$r = 9$$
   The radius of the hemisphere is 9 cm.

ii) Calculate the outer surface area, correct to 1 d.p.

Surface area of hemisphere = area of base
+ area of curved surface

$$= \pi r^2 + \tfrac{1}{2} \times 4\pi r^2$$
$$= 3\pi r^2$$
$$= 3 \times 3.14 \times 9 \times 9$$
$$= 763.0 \text{ to } 1 \text{ d.p.}$$

The outer surface area of the hemisphere is 763.0 cm².

c) A sphere has a surface area of 441π cm². Use π = 3.14 to calculate the following.

i) Calculate the radius.

Surface area of sphere $= 4\pi r^2$
$$4\pi r^2 = 441\pi$$
$$r^2 = 110.25$$
$$r = 10.5$$

The radius of the sphere is 10.5 cm.

ii) Calculate the volume, correct to 1 d.p.

Volume of sphere $= \tfrac{4}{3}\pi r^3$
$$= \tfrac{4}{3} \times 3.14 \times (10.5)^3$$
$$= 4846.6$$

The volume of the sphere is 4846.6 cm³.

d) A solid figure is made from a right circular cone on top of a hemisphere. Both the cone and the hemisphere have a radius of 3 cm. The volume of the hemisphere is $1\tfrac{1}{2}$ times the volume of the cone.

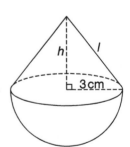

Use π = 3.14 to calculate the following.

i) Calculate the height of the cone.

$1\tfrac{1}{2}$ vol. of cone = vol. of hemisphere
$$\tfrac{3}{2} \times \tfrac{1}{3}\pi r^2 h = \tfrac{1}{2} \times \tfrac{4}{3}\pi r^3$$
$$\tfrac{1}{2}h = \tfrac{2}{3}r$$
$$h = \tfrac{2}{3} \times 2 \times 3$$
$$= 4$$

The height of the cone is 4 cm.

ii) Calculate the total surface area of the solid, correct to 1 d.p.

Use Pythagoras to find $l = \sqrt{3^2 + 4^2} = 5$

Surface area of figure = area of curved surface of cone
+ curved surface area of hemisphere
$$= \pi r l + \tfrac{1}{2} \times 4\pi r^2$$
$$= (3.14 \times 3 \times 5) + (2 \times 3.14 \times 3 \times 3)$$
$$= 47.1 + 56.5$$
$$= 103.6$$

The total surface area of the solid is $103.6\,\text{cm}^2$.

Exercise 5

 1 Use $\pi = 3.14$ to calculate the missing values for each sphere in this table.
Copy and complete the table, giving the radii correct to 1 d.p. and the volumes and surface areas correct to 2 d.p.

	Radius	Diameter	Surface area	Volume
a)		36.0 mm		
b)			113.04 m²	
c)				267.95 cm³
d)	7.0 cm			
e)			28.26 mm²	
f)				859.85 cm³
g)		9.0 cm		
h)			907.46 m²	

 2 Use $\pi = 3.14$ to calculate the missing values for each solid hemisphere in the table.
Copy and complete the table, giving the radii correct to 1 d.p. and the volumes and surface areas correct to 2 d.p.

	Radius	Curved surface area	Total surface area	Volume
a)	5.1 mm			
b)		72.60 m²		
c)				4599.05 cm³
d)	1.3 cm			
e)		3677.82 mm²		

3 216 equal small spheres of plastic are melted together to make one solid plastic ball with a radius of 6 cm.
Calculate the radius of each of the small plastic spheres.

4 I want to paint a sphere with a radius of 9 m. 1 litre of paint will cover 6 m² and costs $7.50. Use $\pi = 3.14$ to calculate the cost of painting this sphere to the nearest dollar.

5 A bucket in the shape of a cylinder has an internal diameter of 22 cm. It contains water to a height of 14 cm. A metal ball with a diameter of 12 cm is put into the bucket.
What is the height (correct to 1 d.p.) of the water now?

6 A salt shaker is the shape of a cylinder with a hemisphere on top.
The diameter is 4.5 cm and the cylinder part has a height of 12 cm.
The salt in the shaker takes up half of the **total** volume.
Work out the height of the salt in the shaker.

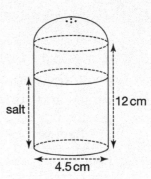

7 A bowl is in the shape of a hemisphere with a cone-shaped part cut out of the middle.
Use $\pi = 3.14$ to calculate the following.

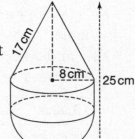

a) How much water does this bowl hold?
b) The bowl is made of a plastic that weighs 2 g for every 1 cm³.
   What is the mass of the bowl to the nearest gram?

8 A solid metal ball with a radius of 6 cm is melted. The metal is used to make a solid cone with a radius of 8 cm. Calculate the height of this cone.

9 A solid figure is made from a hemisphere, a cylinder and a cone joined together as shown in the diagram. All three shapes have the same radius of 8 cm. The total height of the solid is 25 cm. The slant height of the cone is 17 cm. Use $\pi = 3.14$ to calculate:

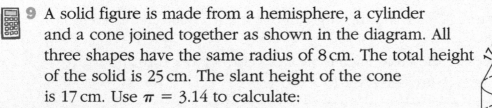

a) the total surface area
b) the volume of the solid, correct to 2 d.p.

Non-linear graphs

Key vocabulary

cubic   parabola   rectangular hyperbola

exponential   quadratic   squared-reciprocal

hyperbola   reciprocal

In Unit 1 we studied equations in the form $y = mx + c$ and found that the graphs of these equations are always straight lines.

During our study of algebra, we have met many other equations that do not have straight lines as their graphs.

To draw the graph of these other equations, we follow the same steps as we did to draw the straight-line graphs.

- Choose values of x.
- Calculate the value of y for each of the x-values you have chosen.
- Choose a scale.
- Plot the points on the Cartesian plane.
- Join the points with a smooth line.
- You can use the graphs to find values of x or y for given values of the other variable, just as before.

NOTE: These graphs are not straight lines so you will not be able to use a ruler to draw the lines. It is possible to buy special 'rulers' that you can bend into curved shapes to help you draw these curved graphs. Ask your teacher about these 'adjustable curve rulers' and other ways you can draw smooth curves.
(Do **not** join each of the plotted points with a short straight line.)

A Non-linear graphs

We will first look at the shape of the graph for each of the main types of equations.

Quadratic graphs

We already know that **quadratic** equations have the form
$y = ax^2 + bx + c$.
We will look at the shape of these graphs when a is positive and when a is negative.

i) $a > 0$

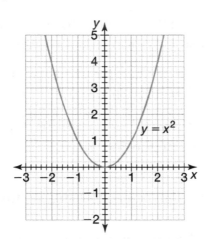

ii) $a < 0$

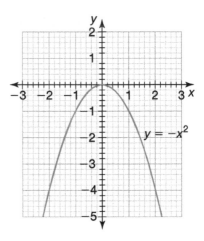

The shape of the graphs of these quadratic equations is known as a parabola.

Cubic graphs

Cubic equations have the general form $y = ax^3 + bx^2 + cx + d$.
We will look at the shape of these graphs when a is positive and when a is negative.

i) $a > 0$

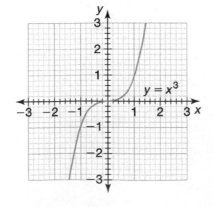

ii) $a < 0$

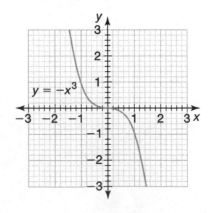

Reciprocal graphs

Reciprocal equations have the general form $y = \dfrac{a}{x}$ where x cannot be 0. We will look at the shape of these graphs when a is positive and when a is negative.

i) $a > 0$

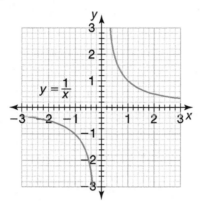

ii) $a < 0$

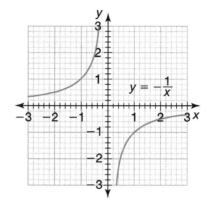

Because x can never be 0, these graphs can never cross the y-axis (where $x = 0$). Do you think these reciprocal graphs can ever cross the x-axis? (Which values of x could make $y = 0$?)

- Graphs of reciprocal equations like this are part of the group of graphs called **hyperbolas**.
- This reciprocal graph is called a **rectangular hyperbola**.
- Even though there are two separate parts of the graph, it is still only one graph and not two separate graphs.
- The two parts of a rectangular hyperbola are mirror images of each other. The equation of the mirror line is either $y = -x$ (when $a > 0$) or $y = x$ (when $a < 0$).

Squared-reciprocal graphs

Squared-reciprocal equations have the general form $y = \dfrac{a}{x^2}$ where x cannot be 0. We will look at the shape of these graphs when a is positive and when a is negative.

i) $a > 0$

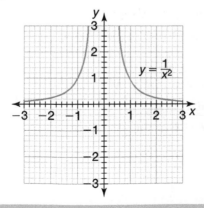

ii) $a < 0$

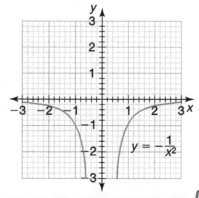

Because x can never be 0, these graphs can never cross the y-axis (where $x = 0$). Do you think these reciprocal graphs can ever cross the x-axis? (Which values of x could make $y = 0$?)

The values of y are always positive when $a > 0$, and they are always negative when $a < 0$. (Can you explain this?)

- The graph of the squared-reciprocal equation is another example of a **hyperbola**.
- Even though there are two separate parts of the graph, it is still only one graph and not two separate graphs.
- The two parts of this hyperbola are also mirror images of each other. This time, the mirror line is the y-axis.

Exponential graphs

Exponential equations have the general form $y = a^x$.
We will look at the shape of these graphs for $y = a^x$ and $y = a^{-x}$ for values of $a > 0$.

i) $a = 2$

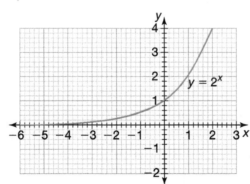

ii) $a = 2$

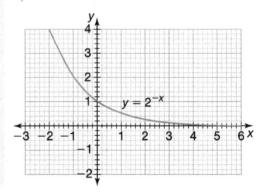

For these equations, the y-values cannot be negative for real number values of x.

Can you think what would happen if $a = 1$ or $0 < a < 1$?

Circle graphs

We have studied circles and their geometrical properties. We can also plot the graph of a circle.

The general equation of a circle with the origin $(0, 0)$ as its centre is in the form $x^2 + y^2 = r^2$, where r is the radius of the circle.

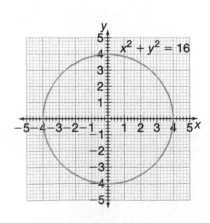

You will need to **recognise the shape** of each of the graphs in this section and know what kind of equation they have.

Example

a) Copy and complete this table of values for the equation $y = x^2 + 3x - 4$.

x	−6	−5	−4	−3	−2	−1	0	1	2
y									

b) Plot these points on a Cartesian plane using a suitable scale, and join them with a smooth curve.

c) Use the graph to find:

i) the value of y when $x = -5.5$ and $x = 1.6$
ii) the value of x when $y = -2$ and $y = 8$.

a) First work out the values of y.

$$y = x^2 + 3x - 4$$
$$= (-6)^2 + 3(-6) - 4$$
$$= 36 - 18 - 4$$
$$= 14$$

$$y = x^2 + 3x - 4$$
$$= (-3)^2 + 3(-3) - 4$$
$$= 9 - 9 - 4$$
$$= -4$$

$$y = x^2 + 3x - 4$$
$$= (0)^2 + 3(0) - 4$$
$$= 0 + 0 - 4$$
$$= -4$$

$$y = x^2 + 3x - 4$$
$$= (-5)^2 + 3(-5) - 4$$
$$= 25 - 15 - 4$$
$$= 6$$

$$y = x^2 + 3x - 4$$
$$= (-2)^2 + 3(-2) - 4$$
$$= 4 - 6 - 4$$
$$= -6$$

$$y = x^2 + 3x - 4$$
$$= (1)^2 + 3(1) - 4$$
$$= 1 + 3 - 4$$
$$= 0$$

$$y = x^2 + 3x - 4$$
$$= (-4)^2 + 3(-4) - 4$$
$$= 16 - 12 - 4$$
$$= 0$$

$$y = x^2 + 3x - 4$$
$$= (-1)^2 + 3(-1) - 4$$
$$= 1 - 3 - 4$$
$$= -6$$

$$y = x^2 + 3x - 4$$
$$= (2)^2 + 3(2) - 4$$
$$= 4 + 6 - 4$$
$$= 6$$

This gives us:

x	−6	−5	−4	−3	−2	−1	0	1	2
y	14	6	0	−4	−6	−6	−4	0	6

b) Choose a scale so that these x- and y-values will fit on your graph paper.

For example, x-axis: 2 cm to stand for 1 unit
y-axis: 1 cm to stand for 1 unit

Plot the points and join them with a smooth curve.

c) Read off the graph:

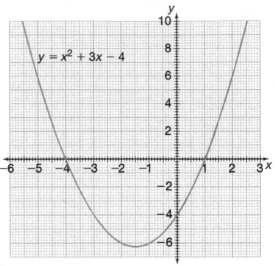

i) when $x = -5.5$, $y = 9.8$
when $x = 1.6$, $y = 3.4$
i.e. the points $(-5.5, 9.8)$
and $(1.6, 3.4)$

ii) when $y = -2$, $x = -3.6$
or 0.6
when $y = 8$, $x = -5.3$ or 2.3
i.e. the points $(-3.6, -2)$ or
$(0.6, -2)$, and $(-5.3, 8)$ or
$(2.3, 8)$

Example

a) Copy and complete this table of values for the equation
$y = x^3 - 5$.

x	−3	−2	−1	0	1	2	3
y							

b) Plot these points on a Cartesian plane using a suitable scale, and join them with a smooth curve.

c) Use your graph to find:

i) the value of y when $x = -2.2$
ii) the value of x when $y = 14$.

a) First work out the values of y.

$$y = x^3 - 5 \qquad y = x^3 - 5 \qquad y = x^3 - 5 \qquad y = x^3 - 5$$
$$= (-3)^3 - 5 \qquad = (-1)^3 - 5 \qquad = (1)^3 - 5 \qquad = (3)^3 - 5$$
$$= -27 - 5 \qquad = -1 - 5 \qquad = 1 - 5 \qquad = 27 - 5$$
$$= -32 \qquad = -6 \qquad = -4 \qquad = 22$$

$$y = x^3 - 5 \qquad y = x^3 - 5 \qquad y = x^3 - 5$$
$$= (-2)^3 - 5 \qquad = (0)^3 - 5 \qquad = (2)^3 - 5$$
$$= -8 - 5 \qquad = 0 - 5 \qquad = 8 - 5$$
$$= -13 \qquad = -5 \qquad = 3$$

This gives us:

x	−3	−2	−1	0	1	2	3
y	−32	−13	−6	−5	−4	3	22

b) Choose a scale so that these x- and y-values will fit on your graph paper.

For example, x-axis: 2 cm to stand for 1 unit

   y-axis: 1 cm to stand for 5 units

Plot the points and join them with a smooth curve.

c) Read off the graph:

   i) when $x = -2.2$, $y = -15.5$
      i.e. the point $(-2.2, -15.5)$
   ii) when $y = 14$, $x = 2.65$
      i.e. the point $(2.65, 14)$

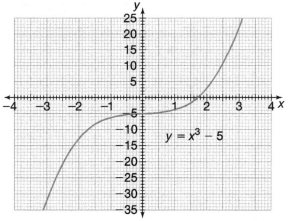

$y = x^3 - 5$

Exercise 1

1 For each equation, copy and complete the table of x- and y-values, and then plot and draw the graph of the equation. Use the graph to read off the given values.

a) $y = 2x^2 - 7x - 9$

x	-2	-1	0	1	2	3	4	5	6
y									

Estimate:
   i) the value(s) of y when $x = -0.5$ and 4.3
   ii) the value(s) of x when $y = -6$ and 8.

b) $y = 4 + 2x - x^2$

x	-3	-2	-1	0	1	2	3	4	5
y									

Estimate:
   i) the value(s) of y when $x = 2.5$
   ii) the value(s) of x when $y = -1$

c) $y = x^2 - 4$

x	-4	-3	-2	-1	0	1	2	3	4
y									

Estimate:
   i) the value(s) of y when $x = -1.5$
   ii) the value(s) of x when $y = 0$.

d) $y = 18x - 5x^2$

x	0	$\frac{1}{2}$	1	$1\frac{1}{2}$	2	$2\frac{1}{2}$	3	$3\frac{1}{2}$	4
y									

Estimate:
  i) the value(s) of y when $x = 1.2$ and 2.8
  ii) the value(s) of x when $y = 1.8$ and 8.8.

2 For each equation, copy and complete the table of x- and y-values, and then plot and draw the graph of the equation. Use the graph to read off the given values.

a) $y = 2x^3 - 9x$

x	−3	−2	−1	0	1	2	3
y							

Estimate:
  i) the value(s) of y when $x = -1.8$ and 2.3
  ii) the value(s) of x when $y = 15$ and -19.

b) $y = x(9 - x^2)$

x	−3	−2	−1	0	1	2	3
y							

Estimate the value(s) of x when $y = 6$.

3 For each equation, copy and complete the table of x- and y-values, and then plot and draw the graph of the equation. Use the graph to read off the given values.

a) $y = 2x + \frac{16}{x} - 11$

x	1	$1\frac{1}{2}$	2	$2\frac{1}{2}$	3	4	5	6
y								

Estimate:
  i) the value(s) of y when $x = 1.7$ and 5.3
  ii) the value(s) of x when $y = 0.8$ and 3.2.

b) $y = -\frac{2}{x} - 1$

x	$\frac{1}{2}$	1	$1\frac{1}{2}$	2	3	4
y						

Estimate:

i) the value(s) of y when $x = 2.5$

ii) the value(s) of x when $y = -1.6$.

c) $y = \frac{4}{x^2}$

x	-3	-2	-1	$-\frac{1}{2}$	$-\frac{1}{4}$	$\frac{1}{4}$	$\frac{1}{2}$	1	2	3
y										

Estimate the value(s) of x when $y = 2$.

4 Copy and complete the table of x- and y-values, and then plot and draw the graph of the equation. Use the graph to read off the given values.

$y = 2 + 2^x$

x	1	$-\frac{1}{2}$	0	1	$1\frac{1}{2}$	2	$2\frac{1}{2}$	3
y								

Estimate:

i) the value(s) of y when $x = -0.7$ and 2.7

ii) the value(s) of x when $y = 5.3$ and 7.5.

B Using graphs to solve quadratic equations

Now we will study the graphs of quadratic equations in more detail to expand what we learned about solving these equations in Unit 3.

One of the ways to solve quadratic equations is to use **factorisation** and the **zero product theorem**.

For example, $\quad 2x^2 + 7x - 4 = 0$

Factorising: $\quad (2x - 1)(x + 4) = 0$

$$2x - 1 = 0 \quad or \quad x + 4 = 0$$
$$x = \frac{1}{2} \quad or \quad \qquad x = -4$$

But not all quadratic expressions will factorise. If the expression does not factorise, we can **complete the square** or use the **quadratic formula**.

We can also use the graph of the quadratic equation to find the values of x when $y = 0$. We draw the graph of $y = 2x^2 + 7x - 4$ and then read off the values of x where the graph crosses the x-axis.

x	$-4\frac{1}{2}$	-4	-3	-2	-1	0	1	$1\frac{1}{2}$
y	5	0	-7	-10	-9	-4	5	11

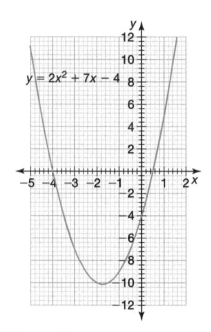

The graph crosses the x-axis at the points $(-4, 0)$ and $\left(\frac{1}{2}, 0\right)$. So the values of x that make the quadratic expression $2x^2 + 7x - 4$ equal to 0 are:

$$x = -4 \quad \text{and} \quad x = \frac{1}{2}$$

These are exactly the same values for x that we found by factorising and using the zero product theorem.

If the quadratic expression does not factorise, we will usually still be able to read the values of x from the graph when $y = 0$. Sometimes the quadratic graph does not cut the x-axis and in this case there are no values of x that will make the quadratic expression equal 0.

C Using graphs to solve simultaneous equations

In Unit 5 we learned how to solve simultaneous linear equations using graphs. We can also use graphs to help us solve simultaneous equations when one of them is a quadratic equation.

Example

Draw the graphs of $y = x^2 - 5x + 6$ and $y = x + 1$. Use these graphs to solve the simultaneous equations $y = x^2 - 5x + 6$ and $y = x + 1$.

First choose values of x and y to plot for each equation.

$y = x^2 - 5x + 6$

x	0	1	2	3	4	5
y	6	2	0	0	2	6

$y = x + 1$

x	0	1	2	3	4	5
y	1	2	3	4	5	6

Now plot these two graphs on the same Cartesian plane.

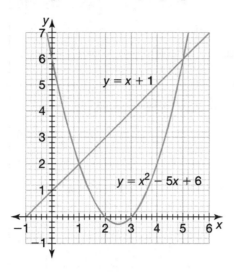

The points where the two graphs cross give us the solutions to the two simultaneous equations.

So the solutions are $x = 1$, $y = 2$ and $x = 5$, $y = 6$.

At these two points, we also can say that:

$x^2 - 5x + 6 = x + 1$

and the solution to this single equation in x is $x = 1$ or $x = 5$.

NOTE: We can also use these graphs to solve the quadratic equation $x^2 - 6x + 5 = 0$. If we simplify $x^2 - 5x + 6 = x + 1$ we get $x^2 - 6x + 5 = 0$ and so they are the same.
So we can say that the solution of the equation $x^2 - 6x + 5 = 0$ is also $x = 1$ or $x = 5$.

Example

Draw the graph for the equation $y = x^2 - x - 3$.
Use the graph to solve the following equations.

a) $x^2 - x - 3 = 0$

If we are drawing the graph of the equation $y = x^2 - x - 3$, we need to decide which other equations to draw so that we can find the points where these graphs cross with the first one. In other words, we are looking for $x^2 - x - 3 = $ 'another equation'.
$x^2 - x - 3 = 0$
The LHS is exactly the RHS of the equation we are drawing, so we will find the points on the graph where $y = 0$ (i.e. where the graph crosses the x-axis).

The graph of
$y = x^2 - x - 3$
crosses the x-axis at the
points $(-1.3, 0)$ and $(2.3, 0)$.

So the solutions to the
equation $x^2 - x - 3 = 0$ are:
$x = -1.3$ or $x = 2.3$

b) $x^2 - x - 9 = 0$

Change the LHS to look
exactly like the RHS of the
equation we are drawing:

$x^2 - x - 9 + 6 = 0 + 6$
or $\qquad x^2 - x - 3 = 6$

So the other line we must
draw is $y = 6$ to see where
the graphs cross.

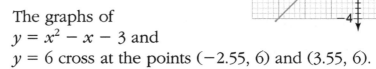

The graphs of
$y = x^2 - x - 3$ and
$y = 6$ cross at the points $(-2.55, 6)$ and $(3.55, 6)$.

So the solutions to the equation $x^2 - x - 3 = 6$ or $x^2 - x - 9 = 0$
are:
$x = -2.55 \quad$ or $\quad x = 3.55$

c) $x^2 - 2x - 2 = 0$

Change the LHS to look exactly like the RHS of the equation we
are drawing:

$x^2 - 2x - 2 + x - 1 = 0 + x - 1$
or $\qquad\qquad x^2 - x - 3 = x - 1$

So the other line we must draw is $y = x - 1$ to see where the
graphs cross.

The graphs of $y = x^2 - x - 3$ and $y = x - 1$ cross at the points
$(-0.75, -1.75)$ and $(2.75, 1.75)$.

So the solutions to the equation $x^2 - x - 3 = x - 1$
or $x^2 - 2x - 2 = 0$ are:
$x = -0.75$ or $x = 2.75$

Exercise 2

1 Imagine you have drawn the graph of the equation $y = 2x^2 + 3$. Work out the equation of a **straight line** that you must draw to help you solve each of these equations.

a) $2x^2 + 5x + 1 = 0$ b) $2x^2 - 2x - 1 = 0$
c) $x^2 - 3x - 1 = 0$ d) $4x^2 + 8x - 5 = 0$

2 a) Copy and complete the table of values for x and y, and plot and draw the graph.

$y = x^2 - x - 1$

x	−2	−1	0	1	2	3
y						

b) Use this graph to solve the equation $x^2 - x - 1 = 0$.

3 a) Copy and complete the table of values for x and y, and plot and drow the graph.

$y = x^2 - 6x + 7$

x	0	1	2	3	4	5	6
y							

b) Use this graph to solve the equation $x^2 - 6x + 7 = 0$.

4 a) Copy and complete the table of values for x and y, and plot and draw the graph.

$y = x^2 - x - 4$

x	−3	−2	−1	0	1	2	3	4
y								

b) Use this graph to solve these equations:
   i) $x^2 - x - 4 = 2$
   ii) $x^2 - x - 4 = -3$
   iii) $x^2 - x - 8 = 0$

5 a) Copy and complete the table of values for x and y, and plot and draw the graph.

$y = 4x^2 - 8x - 5$

x	−2	−1	0	1	2	3	4
y							

b) Use this graph to solve these equations:
   i) $4x^2 - 8x = 0$
   ii) $4x^2 - 8x + 4 = 0$
   iii) $4x^2 - 8x - 21 = 0$

6 a) Copy and complete the table of values for x and y, and plot and draw the graph.

$y = 5x - 2x^2 + 2$

x	−1	$-\frac{1}{2}$	0	$\frac{1}{2}$	1	$1\frac{1}{2}$	2	$2\frac{1}{2}$	3
y									

b) Use this graph to solve the equations:
   i) $5x - 2x^2 + 2 = 0$
   ii) $5x - 2x^2 - 1 = 0$

7 a) Copy and complete the table of values for x and y, and plot and draw the graph.

$y = x^3 - 6x$

x	−3	−2	−1	0	1	2	3
y							

b) Use this graph to solve the equations:
   i) $x^3 - 6x + 6 = 0$
   ii) $x^3 - 6x - 2 = 0$

Unit 10

An introduction to trigonometry

angle of depression	cosine	surveying
angle of elevation	hypotenuse	tangent
astronomy	navigation	trigonometric ratio
bearing	opposite	trigonometry
compass	sine	adjacent

The word trigonometry comes from two Greek words *trigon*, which means 'triangle', and the word *metreo*, which means 'to measure'. So trigonometry is the study of measuring triangles.

In Unit 10 of Coursebook 2 we learned about the theorem of Pythagoras. This theorem gives a link between the lengths of the three sides of a right-angled triangle.
Pythagoras' theorem tells us that the length of the longest side of a right-angled triangle (called the hypotenuse), c, is connected to the lengths of the two shorter sides, a and b, with this equation:

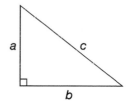

$$c^2 = a^2 + b^2$$

Now we are going to learn more about right-angled triangles and how the lengths of their sides and their angles are connected.
Before we do that, it will be useful to give a name to each of the shorter sides of a right-angled triangle in order to distinguish between them.
The most important angle in these triangles is the right angle.
The most important side (and the longest) we already know is called the **hypotenuse**. This is the side that is **opposite the right angle** in the triangle.

Now let's give a name to the other two sides in any right-angled triangle: Look at $\triangle ABC$. We can see that $\angle C$ is the right angle and that side c (AB) is opposite this angle.
We know this side is called the hypotenuse.
We name the other two sides depending on which **angle** we are looking at.

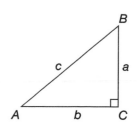

For ∠A: Side *a* (*BC*) is opposite ∠A and this side is called the opposite side for ∠A. Side *b* (*AC*) is next to ∠A and this side is called the adjacent side for ∠A (adjacent means 'next to').

For ∠B: Side *b* (*AC*) is opposite ∠B and this side is called the **opposite side** for ∠B. Side *a* (*BC*) is next to ∠B and this side is called the **adjacent side** for ∠B.

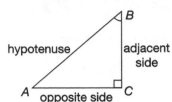

Trigonometric ratios

We know how the lengths of the sides in a right-angled triangle are connected. Now we are going to connect the lengths of the sides with the sizes of the other two angles in the triangle.

Here are the new rules:

For ∠A: $\dfrac{\text{length of opposite side}}{\text{length of hypotenuse}}$ $\left(\text{or } \dfrac{BC}{AB}\right)$ is called the sine of ∠A.

$\dfrac{\text{length of adjacent side}}{\text{length of hypotenuse}}$ $\left(\text{or } \dfrac{AC}{AB}\right)$ is called the cosine of ∠A.

$\dfrac{\text{length of opposite side}}{\text{length of adjacent side}}$ $\left(\text{or } \dfrac{BC}{AC}\right)$ is called the tangent of ∠A.

For ∠B: $\dfrac{\text{length of opposite side}}{\text{length of hypotenuse}}$ $\left(\text{or } \dfrac{AC}{AB}\right)$ is called the **sine** of ∠B.

$\dfrac{\text{length of adjacent side}}{\text{length of hypotenuse}}$ $\left(\text{or } \dfrac{BC}{AB}\right)$ is called the **cosine** of ∠B.

$\dfrac{\text{length of opposite side}}{\text{length of adjacent side}}$ $\left(\text{or } \dfrac{AC}{BC}\right)$ is called the **tangent** of ∠B.

NOTE: **a)** It doesn't matter how big or small the triangle is, these trigonometric ratios of the lengths of the sides will always be the same for the same size of the angles.

b) The value of sine, cosine and tangent for any angle is just a number with **no units** (because it is one length divided by another length).

So, for **any** angle x (except the right angle) in **any** right-angled triangle:

Rule	Usual abbreviation
sine $x = \dfrac{\text{opposite side}}{\text{hypotenuse}}$	$\sin x = \dfrac{\text{opp}}{\text{hyp}}$
cosine $x = \dfrac{\text{adjacent side}}{\text{hypotenuse}}$	$\cos x = \dfrac{\text{adj}}{\text{hyp}}$
tangent $x = \dfrac{\text{opposite side}}{\text{adjacent side}}$	$\tan x = \dfrac{\text{opp}}{\text{adj}}$

You will need to learn these rules and remember them.
There are several funny sentences or expressions to help us remember these three rules. Here is one of them:

'some old hens cackle and hover till old age'

Compare it with the rules:

$$\sin x = \frac{\text{opp}}{\text{hyp}} \qquad \cos x = \frac{\text{adj}}{\text{hyp}} \qquad \tan x = \frac{\text{opp}}{\text{adj}}$$

Can you see the pattern that will help you to remember the rules?

Exercise 1

1 For each of these right-angled triangles, write down the name of:

  i) the hypotenuse   ii) the side adjacent to $\angle f$

  iii) the side opposite $\angle f$.

a)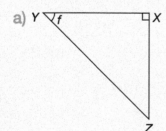

b)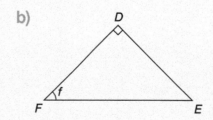

c)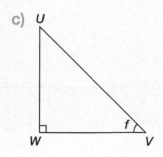

d)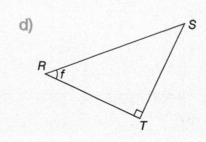

2 For each of these right-angled triangles, write down an expression for:

　i) $\sin D$　　　　ii) $\cos D$　　　　iii) $\tan D$
　iv) $\sin E$　　　　v) $\cos E$　　　　vi) $\tan E$

a)
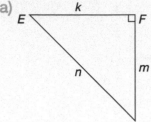

b)
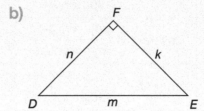

c)
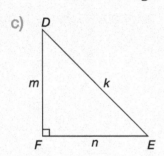

d)
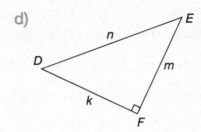

3 For each of these right-angled triangles (not drawn to scale), calculate a value for:

　i) $\sin P$　　　　ii) $\cos P$　　　　iii) $\tan P$
　iv) $\sin Q$　　　　v) $\cos Q$　　　　vi) $\tan Q$

Give your answers correct to 4 decimal places.

a)
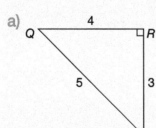

b)
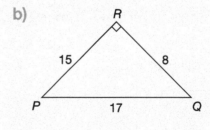

c)
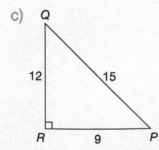

d)
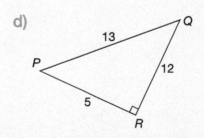

What do you notice about the answers for parts **a)** and **c)**? Can you think of a reason for this?

B The values of the trigonometric ratios

We know that, for any given angle x, the values of $\sin x$, $\cos x$ and $\tan x$ are constant. For example, $\sin 30° = 0.5000$ for every right-angled triangle, no matter how big or small:

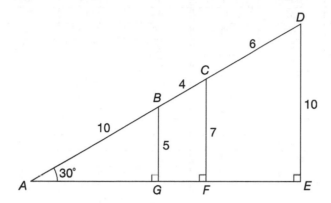

In the diagram, $\angle A = 30°$.

In $\triangle ABG$ $\sin A = \dfrac{\text{opp}}{\text{hyp}} = \dfrac{BG}{AB} = \dfrac{5}{10} = 0.5000$

In $\triangle ACF$ $\sin A = \dfrac{\text{opp}}{\text{hyp}} = \dfrac{CF}{AC} = \dfrac{7}{10 + 4} = 0.5000$

In $\triangle ADE$ $\sin A = \dfrac{\text{opp}}{\text{hyp}} = \dfrac{DE}{AD} = \dfrac{10}{10 + 4 + 6} = 0.5000$

The value of $\cos A$ will also be constant for all three triangles ($\cos 30° = 0.8660$). The value of $\tan A$ will also be constant for all three triangles ($\tan 30° = 0.5774$).

Because the values of sin, cos and tan of each angle never change, it is possible to make a table of all these values for the common angles so that we can look them up when we need them.

You can also use a calculator to find out the value of sin, cos or tan of any angle. Check your calculator to see if it is one that allows you to calculate these trigonometric values. All calculators are different and the order and type of the keys you must press to get the right answers will be slightly different. If you don't know how to work out these values on the calculator that you have, ask your teacher to show you.

If you don't have a calculator, or yours doesn't allow you to calculate these trigonometric values, you can use a table of the values of sin, cos and tan of the angles from 0° to 90° (see page 168):

A table of trigonometric ratios

Degree	Sin	Cos	Tan	Degree	Degree	Sin	Cos	Tan	Degree
00	.0000	1.0000	.0000	00	46	.7193	.6947	1.0355	46
01	.0175	.9998	.0175	01	47	.7314	.6820	1.0724	47
02	.0349	.9994	.0349	02	48	.7431	.6691	1.1106	48
03	.0523	.9986	.0524	03	49	.7547	.6561	1.1504	49
04	.0698	.9976	.0699	04	50	.7660	.6428	1.1918	50
05	.0872	.9962	.0875	05	51	.7771	.6293	1.2349	51
06	.1045	.9945	.1051	06	52	.7880	.6157	1.2799	52
07	.1219	.9925	.1228	07	53	.7986	.6018	1.3270	53
08	.1392	.9903	.1405	08	54	.8090	.5878	1.3764	54
09	.1564	.9877	.1584	09	55	.8192	.5736	1.4281	55
10	.1736	.9848	.1763	10	56	.8290	.5592	1.4826	56
11	.1908	.9816	.1944	11	57	.8387	.5446	1.5399	57
12	.2079	.9781	.2126	12	58	.8480	.5299	1.6003	58
13	.2250	.9744	.2309	13	59	.8572	.5150	1.6643	59
14	.2419	.9703	.2493	14	60	.8660	.5000	1.7321	60
15	.2588	.9659	.2679	15	61	.8746	.4848	1.8040	61
16	.2756	.9613	.2867	16	62	.8829	.4695	1.8807	62
17	.2924	.9563	.3057	17	63	.8910	.4540	1.9626	63
18	.3090	.9511	.3249	18	64	.8988	.4384	2.0503	64
19	.3256	.9455	.3443	19	65	.9063	.4226	2.1445	65
20	.3420	.9397	.3640	20	66	.9135	.4067	2.2460	66
21	.3584	.9336	.3839	21	67	.9205	.3907	2.3559	67
22	.3746	.9272	.4040	22	68	.9272	.3746	2.4751	68
23	.3907	.9205	.4245	23	69	.9336	.3584	2.6051	69
24	.4067	.9135	.4452	24	70	.9397	.3420	2.7475	70
25	.4226	.9063	.4663	25	71	.9455	.3256	2.9042	71
26	.4384	.8988	.4877	26	72	.9511	.3090	3.0777	72
27	.4540	.8910	.5095	27	73	.9563	.2924	3.2709	73
28	.4695	.8829	.5317	28	74	.9613	.2756	3.4874	74
29	.4848	.8746	.5543	29	75	.9659	.2588	3.7321	75
30	.5000	.8660	.5774	30	76	.9703	.2419	4.0108	76
31	.5150	.8572	.6009	31	77	.9744	.2250	4.3315	77
32	.5299	.8480	.6249	32	78	.9781	.2079	4.7046	78
33	.5446	.8387	.6494	33	79	.9816	.1908	5.1446	79
34	.5592	.8290	.6745	34	80	.9848	.1736	5.6713	80
35	.5736	.8192	.7002	35	81	.9877	.1564	6.3138	81
36	.5878	.8090	.7265	36	82	.9903	.1392	7.1154	82
37	.6018	.7986	.7536	37	83	.9925	.1219	8.1443	83
38	.6157	.7880	.7813	38	84	.9945	.1045	9.5144	84
39	.6293	.7771	.8098	39	85	.9962	.0872	11.4301	85
40	.6428	.7660	.8391	40	86	.9976	.0698	14.3007	86
41	.6561	.7547	.8693	41	87	.9986	.0523	19.0811	87
42	.6691	.7431	.9004	42	88	.9994	.0349	28.6363	88
43	.6820	.7314	.9325	43	89	.9998	.0175	57.2900	89
44	.6947	.7193	.9657	44	90	1.0000	.0000	Infinity	90
45	.7071	.7071	1.0000	45					

In a right-angled triangle, the hypotenuse is always the longest side. This means that the length of the hypotenuse is always bigger than the length of the adjacent side or the opposite side for any angle. Think about the values of sine and cosine for an angle:

$$\sin x = \frac{\text{opp}}{\text{hyp}} \quad \text{and} \quad \cos x = \frac{\text{adj}}{\text{hyp}}$$

Because the hypotenuse is always bigger, the values of sine and cosine for any angle are always **less than 1**.

The value of tangent is not restricted in this way $\left(\tan x = \dfrac{\text{opp}}{\text{adj}}\right)$ and we find values of any size for the tangent of different angles. Look at the table on page 168 to confirm this.

NOTE: In the calculations that follow in this unit, the sizes of some of the angles are not whole numbers. If you can use a calculator to find the values of sine, cosine and tangent, then this will not be a problem. If you are using the table of values supplied, then you will need to round the values of the angles to the **nearest whole number** before looking up the trigonometric values. If you are using the table to look up the angle from a trigonometric value, then find the angle that has a trigonometric value **closest** to the one you are given.

The following exercise will give you practice in using a calculator and/or tables to find the trigonometric values for angles, and vice versa.

 Exercise 2

 1 Use a calculator or tables to find these trigonometric ratios.
Give the answers correct to 4 decimal places.

a) $\tan 40°$	b) $\sin 30°$	c) $\cos 0°$
d) $\tan 46°$	e) $\cos 60°$	f) $\sin 58°$
g) $\tan 35°$	h) $\sin 90°$	i) $\cos 12.3°$
j) $\tan 12.6°$	k) $\cos 87.9°$	l) $\sin 75.3°$
m) $\tan 64.31°$	n) $\cos 88.36°$	o) $\sin 77.45°$

2 Use a calculator and/or tables to evaluate each of these.
Give answers correct to 3 decimal places.

a) $\cos 28° + \cos 53°$ b) $8 \tan 21° - 4 \sin 12°$ c) $13 \cos 14° \times 11 \tan 48°$

d) $9 \sin 17° \div 2 \cos 48°$ e) $4 \cos 87° + \sin 3°$ f) $13 \tan 15.8° - 2 \sin 14.6°$

g) $3 \cos 43.4° + 2 \sin 15.7°$ h) $2 \tan 17.8° \times \sin 70.5°$ i) $5 \cos 69.8° \div 2 \sin 30.7°$

 3 If y is an angle between $0°$ and $90°$, find the value of y for each of these
trigonometric ratios, using a calculator or tables. Give the answers correct to
1 decimal place.

a) $\tan y = 0.3$ b) $\sin y = 0.4$ c) $\cos y = 1$

d) $\tan y = 12$ e) $\cos y = 0.74$ f) $\sin y = 0.46$

g) $\tan y = 1.34$ h) $\sin y = 0.317$ i) $\cos y = 0.973$

j) $\tan y = 4.371$ k) $\cos y = 0.707$ l) $\sin y = 0.1736$

m) $\tan y = 3.145$ n) $\cos y = 0.987$ o) $\sin y = 1$

 4 $\triangle ABC$ is a right-angled triangle as shown in the diagram.

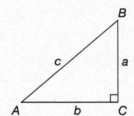

Use this diagram with a calculator or tables to copy and
complete each of these statements to make them true.

a) If $\angle A = 20°$, then $\dfrac{a}{c} =$ _____   b) If $\dfrac{a}{c} = 0.5736$, then $\angle A =$ _____

c) If $\angle A = 75°$, then $\dfrac{b}{c} =$ _____   d) If $\dfrac{b}{c} = 0.5000$, then $\angle A =$ _____

e) If $\angle A = 55°$, then $\dfrac{a}{b} =$ _____   f) If $\dfrac{a}{b} = 0.8391$, then $\angle A =$ _____

g) If $\angle B = 5°$, then $\dfrac{a}{c} =$ _____   h) If $\dfrac{a}{c} = 0.4226$, then $\angle B =$ _____

i) If $\angle B = 30°$, then $\dfrac{b}{a} =$ _____   j) If $\dfrac{b}{a} = 5.6713$, then $\angle B =$ _____

k) If $\angle B = 50°$, then $\dfrac{b}{c} =$ _____   l) If $\dfrac{b}{c} = 0.2588$, then $\angle B =$ _____

Finding the unknown sides of a triangle using trigonometric ratios

We can use Pythagoras' theorem to find the length of an unknown side in a right-angled triangle, but we need the lengths of the other two sides to do so. If we are given the length of only one side and the size of one of the angles, we can use the trigonometric ratios to calculate the lengths of the other two sides.

Examples

Find the length of each of the unknown sides in these triangles. Give the answers correct to 2 decimal places.

a)

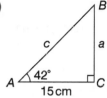

Choose which of the trigonometric ratios to use.

$$\tan 42° = \frac{a}{15}$$

Use a calculator or tables to evaluate $\tan 42°$.

$$0.9004 = \frac{a}{15}$$
$$a = 0.9004 \times 15$$
$$= 13.51$$

So BC is 13.51 cm.

$$\cos 42° = \frac{15}{c}$$

Use a calculator or tables to evaluate $\cos 42°$.

$$0.7431 = \frac{15}{c}$$
$$c = 15 \div 0.7431$$
$$= 20.19$$

So AB is 20.19 cm.

$$(\angle B = 180° - (90° + 42°) = 48°)$$

b)

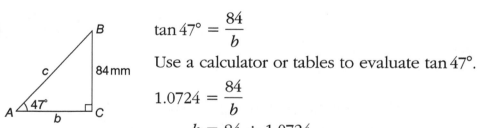

$$\tan 47° = \frac{84}{b}$$

Use a calculator or tables to evaluate $\tan 47°$.

$$1.0724 = \frac{84}{b}$$
$$b = 84 \div 1.0724$$
$$= 78.33$$

So AC is 78.33 mm.

$$\sin 47° = \frac{84}{c}$$

Use a calculator or tables to evaluate $\sin 47°$:
$$0.7314 = \frac{84}{c}$$
$$c = 84 \div 0.7314$$
$$= 114.85$$
So AB is 114.85 mm

$(\angle B = 180° - (90° + 47°) = 43°)$

c)

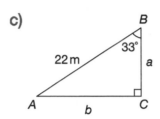

$\cos 33° = \dfrac{a}{22}$

Use a calculator or tables to evaluate $\cos 33°$:

$$0.8387 = \frac{a}{15}$$
$$a = 0.8387 \times 15$$
$$= 12.58$$

So BC is 12.58 m.
$$\sin 33° = \frac{b}{22}$$

Use a calculator or tables to evaluate $\sin 33°$:

$$0.5446 = \frac{b}{22}$$
$$b = 0.5446 \times 22$$
$$= 11.98$$

So AC is 11.98 m.

$(\angle A = 180° - (90° + 33°) = 57°)$

NOTE: You can also check your answers by using Pythagoras' theorem.

 Exercise 3

1 $\triangle ABC$ is a right-angled triangle as shown in the diagram. Use this diagram with a calculator or tables to calculate each of the missing values in the table. Give the lengths of the sides correct to 2 decimal places.

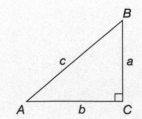

	∠A	∠B	∠C	a	b	c
a)	40°	i)	90°	ii)	3 cm	iii)
b)	34°	i)	90°	ii)	iii)	15 cm
c)	i)	39°	90°	ii)	iii)	8 m
d)	i)	40°	90°	ii)	5 cm	iii)
e)	30°	i)	90°	5 cm	ii)	iii)
f)	51.7°	i)	90°	7.53 m	ii)	iii)
g)	i)	36.5°	90°	ii)	iii)	12 cm
h)	49.6°	i)	90°	ii)	2.7 cm	iii)
i)	i)	19.9°	90°	ii)	iii)	17.9 cm
j)	24°	i)	90°	ii)	iii)	6.5 cm
k)	i)	27.9°	90°	ii)	6.9 cm	iii)
l)	i)	21.6°	90°	ii)	iii)	86.5 m
m)	11.5°	i)	90°	1.5 cm	ii)	iii)

2 △*ABC* is a right-angled triangle as shown in the diagram.
Use this diagram with a calculator or tables to calculate
each of the missing values in the table.
Give the lengths of the sides correct to 2 decimal places.

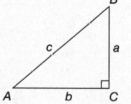

	∠A	∠B	∠C	a	b	c
a)	32°	i)	90°	ii)	27 mm	iii)
b)	i)	23°	90°	ii)	iii)	14 cm
c)	i)	21°	90°	ii)	6 mm	iii)
d)	31°	i)	90°	15.5 m	ii)	iii)
e)	15°	i)	90°	ii)	iii)	37.5 cm
f)	i)	28°	90°	14 mm	ii)	iii)
g)	44.2°	i)	90°	ii)	7 m	iii)
h)	i)	34°	90°	ii)	iii)	12 cm
i)	i)	21.5°	90°	ii)	8.9 cm	iii)
j)	34.3°	i)	90°	ii)	15.7 mm	iii)
k)	69.4°	i)	90°	ii)	iii)	75.6 m
l)	57.6°	i)	90°	14.78 cm	ii)	iii)
m)	i)	23.42°	90°	67.4 cm	ii)	iii)

D Finding the size of an angle using trigonometric ratios

If we know the lengths of two of the sides in a right-angled triangle, it is possible to use the trigonometric ratios to work out the size of the two smaller angles in the triangle. We can also use Pythagoras' theorem to work out the length of the third side.

Examples

Find the size of each of the unknown marked angles and the remaining side in these triangles. Give the answers correct to 4 significant figures.

a)

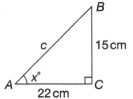

Choose which of the trigonometric ratios to use.

$$\tan x° = \frac{15}{22}$$

$$= 0.6818$$

Use a calculator or tables to find the angle whose tangent is 0.6818.

$$x = 34.29$$

So $\angle A$ is 34.29°.

$$\sin 34.29° = \frac{15}{c}$$

Use a calculator or tables to evaluate $\sin 34.29°$.

$$0.5634 = \frac{15}{c}$$

$$c = 15 \div 0.5634$$

$$= 26.62$$

So AB is 26.62 cm.

b)

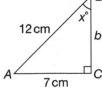

Choose which of the trigonometric ratios to use.

$$\sin x° = \frac{7}{12}$$

$$= 0.5833$$

Use a calculator or tables to find the angle whose sine is 0.5833:

$$x = 35.69$$

So $\angle B$ is 35.69°.

$$\cos 35.69° = \frac{b}{12}$$

Use a calculator or tables to evaluate $\cos 35.69°$.

$$0.8122 = \frac{b}{12}$$

$$b = 0.8122 \times 12 = 9.746$$

So BC is 9.746 cm.

c)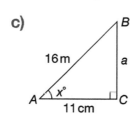

$$\cos x° = \frac{11}{16}$$
$$= 0.6875$$

Use a calculator or tables to find the angle whose cosine is 0.6875.

$$x = 46.57$$

So $\angle A$ is 46.57°.

$$\tan 46.57° = \frac{a}{11}$$

Use a calculator or tables to evaluate $\tan 46.57°$.

$$1.0564 = \frac{a}{11}$$

$$a = 1.0564 \times 11 = 11.62$$

So BC is 11.62 cm.

Exercise 4

1 $\triangle ABC$ is a right-angled triangle as shown in the diagram. Use this diagram with a calculator or tables to calculate each of the missing values in the table. Give the angles correct to 1 d.p. and the sides correct to 2 d.p.

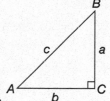

	$\angle A$	$\angle B$	$\angle C$	a	b	c
a)	i)		90°	12 cm	13 cm	ii)
b)		i)	90°	ii)	12.5 m	26 m
c)	i)		90°	15 mm	ii)	22.7 mm
d)		i)	90°	3.2 cm	3.7 cm	ii)
e)		i)	90°	4.6 cm	ii)	5.9 cm
f)	i)		90°	14.7 m	12.5 m	ii)
g)		i)	90°	2.5 cm	ii)	6.4 cm
h)	i)		90°	ii)	3 m	4.7 m
i)	i)		90°	32.8 mm	ii)	41.6 mm

We will now apply what we have learned about trigonometric ratios to help us find the unknown sides and angles in more complex shapes.

Example

Use the trigonometric ratios to calculate the unknown angles, x and y, and unknown side, z, in the diagram.

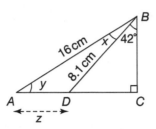

i) Find the length of BC:

In $\triangle BCD$, $\cos 42° = \dfrac{BC}{BD}$

$BC = 0.7431 \times 8.1 = 6.02$

So BC is 6.02 cm.

ii) Find the length of CD:

In $\triangle BCD$, $\sin 42° = \dfrac{CD}{BD}$

$CD = 0.6691 \times 8.1 = 5.42$

So CD is 5.42 cm.

iii) Find angle $A\ (= y)$:

In $\triangle ABC$,

$\sin A = \dfrac{BC}{AB} = \dfrac{6.02}{16} = 0.3763$

So $\angle A$ is 22.1°.

iv) Find angle x:

In $\triangle ABC$, $x + y + 42 + 90 = 180$

$x = 180 - (90 + 42 + 22.1) = 25.9$

So $\angle x$ is 25.9°.

v) Find the length of AC:

In $\triangle ABC$, $\cos A = \dfrac{AC}{AB}$

$AC = 0.9265 \times 16 = 14.82$

So AC is 14.82 cm.

vi) Find the length of $AD\ (= z)$:

$AD = AC - DC$

$= 14.82 - 5.42 = 9.40$

So AD is 9.40 cm.

So $\angle x = 25.9°$, $\angle y = 22.1°$ and $z = 9.40$ cm

 Exercise 5

1 Use the trigonometric ratios to calculate the unknown angles and/or sides in each figure.

Give the angles correct to 1 decimal place and the sides correct to 2 decimal places.

a)

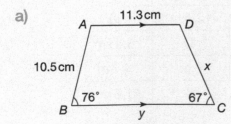

b)

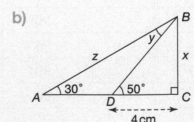

c)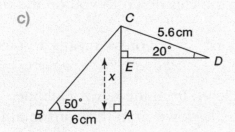

d)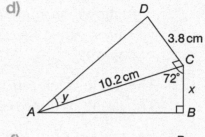

e)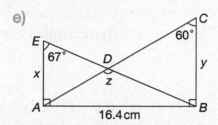

f)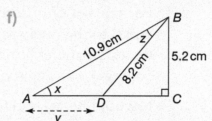

E Practical applications of trigonometry

Here are some of the important fields of study that are based on calculations involving the trigonometric ratios:

- **Surveying** – finding the size and distance of large objects when it is not possible to measure them directly
- **Navigation** – finding the right direction and distance to travel at sea or in space
- **Astronomy** – studying planets, stars and other objects in space and measuring their movements relative to earth and relative to each other.

We will now look at some simple applications of how the trigonometric ratios are used to calculate measurements like these in everyday life.

Angles of elevation and depression

Imagine you are standing on a beach at the bottom of a high cliff. You are looking up at your friend who is standing at the top of the cliff, and your friend is looking down at you.

The line that joins you and your friend makes two special angles:

1 The angle from you on the beach **up** to your friend on the cliff is shown as angle e.
This is called the angle of elevation ('elevate' means 'to raise up').

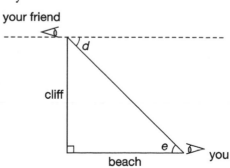

2 The angle from your friend on the cliff **down** to you on the beach is shown as angle d.
This is called the **angle of depression** ('depress' means 'to push down').

The angle of **elevation** is the angle **up** from the horizontal line.
The angle of **depression** is the angle **down** from the horizontal line.

For one set of observations like this, the angle of elevation is always **equal** to the angle of depression. (They are **alternate angles** between the two parallel horizontal lines.)
So angle e = angle d.

It would be quite difficult to measure how high the cliff is, but you could measure how far from the bottom of the cliff you are standing. If you also know the angle of elevation, e, then you can use the trigonometric ratio

$$\tan e = \frac{\text{height of cliff}}{\text{horizontal distance from you to cliff}}$$

to work out the height of the cliff.

Examples

a) The angle of elevation of the top of a cliff from a ship out to sea is 15.7°. The ship is 3.1 km from the cliff.
Calculate the height of the cliff in metres.

Draw a diagram of the information that is given.

The ship is at point S, the bottom of the cliff is at point B, and the top of the cliff is at point C.

$$\tan 15.7° = \frac{CB}{3.1}$$
$$CB = 0.2811 \times 3.1$$
$$= 0.8714$$

So the height of the cliff is 0.8714 km or 871.4 m.

b) Some tourists are at a viewing point 85 m above the ground at the top of the Statue of Liberty in New York. They can see the Staten Island Ferry in the bay, 340 m away from the Statue. Calculate the angle of depression to the ferry from the viewing point.

Draw a diagram of the information that is given.

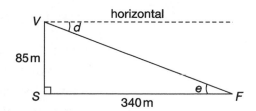

The viewing point is at V and the ferry is at F.

$$\tan e = \frac{85}{340} = 0.2500$$

$$e = 14.036 \ldots$$

The angle of elevation, e, is equal to the angle of depression, so the angle of depression is $14.0°$ (to 1 d.p.).

Exercise 6

 Give the answers correct to 2 decimal places for lengths and correct to 1 decimal place for angles.

1 The angle of elevation from a point 21.5 m from the base of a flagpole is 47°. Calculate the height of the flagpole.

2 A bird is sitting at the top of a tree that is 11.6 m high. A worm is 25.6 m away from the base of the tree on the horizontal ground.
What is the angle of depression from the bird to the worm?

3 The spire on a church is 77 m high. On a sunny afternoon, the shadow of the spire is 53 m long.
Calculate the angle of elevation of the sun.

4 A train is travelling downhill. The altitude of the train has decreased by 19.5 m after travelling half a kilometre on the track.
Calculate the angle of depression of the train.

5 A ladder is 7.2 m long. It is leaning against the wall of a house and touches a window sill. The angle of elevation of the ladder is 63.4°.
Calculate:

a) the height of the window sill above the ground
b) how far the foot of the ladder is from the house.

6 A cliff is 41 m high. There are two boats out to sea in a straight line from the cliff. The angles of depression to the two boats are 15° and 23°.
Calculate the distance between the two boats.

7 A lighthouse stands at the top of a cliff. A dolphin is swimming 63 m away from the cliff. The angle of elevation from the dolphin to the top of the cliff is 39° and to the top of the lighthouse is 49°.
Calculate the height of the lighthouse.

Navigation using three-figure bearings

To find our way, we use a compass.
The four main directions on a compass are
shown as: North, South, East, West.
Together, all the directions on the compass
make up a full circle of 360°. The direction
North is given a value of 0° (also 360°). **East** is
given a value of 90°. **South** is given a value of
180° and **West** is given a value of 270°.

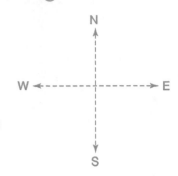

Bearings are used to describe the direction on the compass that you
must travel to get from one place to another.

- A bearing is written as a number of degrees. It can be
  any angle from 0° to 360°.
- A bearing is always measured from the north line (0°),
  turning in a **clockwise** direction.
- A bearing is always written with **three figures**, so if the
  direction is 37° the bearing would be written 037°.

Here are some examples.

P is on a bearing of 059° from *Q*.

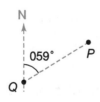

R is on a bearing of 288° from *S*.

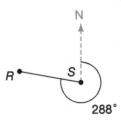

Example

A yacht sails on a bearing of 127° from the harbour for 407 km.
How far south of the harbour is the yacht now?

Draw a diagram of the information given.
The harbour is at *H* and the yacht has sailed to
the point *Y*.
HS is the distance of the yacht directly south of
the harbour.
$\angle YHS = 180° - 127° = 53°$
Now use a trigonometric ratio.

$$\cos 53° = \frac{HS}{407}$$

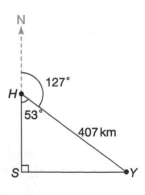

$HS = \cos 53° \times 407 = 0.6018 \times 407 = 244.93$

So the yacht is now 244.93 km directly south of the harbour.

Example

A plane flies 15 km on a bearing of 061° from the airport.
The pilot then changes course and flies for 23 km on a bearing of 152°.
On what bearing must the plane now fly to return directly to the airport?

Draw a diagram of the information given.
The airport is at A. AC is the first part of the flight.
CP is the second part of the flight. PA is the flight home on a bearing of $x°$,
$x° = 270° + \angle APE$
To find $\angle APE$ we need two sides of $\triangle APE$.
We are able to find AE and EP.

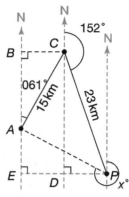

i) $AE = CD - AB$

$$\cos \angle DCP = \frac{CD}{CP}$$

and $\angle DCP = 180° - 152° = 28°$

$$CD = \cos 28° \times 23 = 0.8829 \times 23 = 20.31$$

Also, $\cos \angle BAC = \frac{AB}{AC}$ and $\angle BAC = 61°$

$AB = \cos 61° \times 15 = 0.4848 \times 15 = 7.27$
So $AE = CD - AB = 20.31 - 7.27 = 13.04$
AE is 13.04 km.

ii) $EP = BC + DP$

$$\sin 61° = \frac{BC}{AC}$$

$$BC = \sin 61° \times 15 = 0.8746 \times 15 = 13.12$$

Also, $\sin \angle DCP = \frac{DP}{CP}$

and $\angle DCP = 180° - 152° = 28°$

$$DP = \sin 28° \times 23 = 0.4695 \times 23 = 10.80$$

So, $EP = BC + DP = 13.12 + 10.80 = 23.92$
EP is 23.92 km.

iii) Now, in $\triangle APE$:

$$\tan \angle APE = \frac{AE}{EP} = \frac{13.04}{23.92} = 0.5452$$

$$\angle APE = 28.6°$$

iv) Finally, $x° = 270° + \angle APE$
$= 270° + 28.6° = 298.6°$

The plane must travel on a bearing of 299° to return to the airport.

NOTE: This second example is quite challenging. It illustrates the kind of calculations the navigator in a plane needs to make to reach the correct destination. More advanced students should be able to complete similar calculations.

If you continue to study mathematics, you will learn other trigonometric rules that can be used to solve the problem in the second example in other ways.

Exercise 7

 1 A ship sails out of a harbour for 193 km on a bearing of 058°.

  a) How far north of the harbour is the ship now?
  b) How far east of the harbour is the ship now?

 2 Yacht C is 5.7 km due south of a buoy. Yacht D is 7.9 km due east of the same buoy.

  a) What is the bearing of yacht D from yacht C?
  b) What is the bearing of yacht C from yacht D?

 3 A plane flies from the airport on a bearing of 329°. The plane is now 94 km due west of the airport.
  How far has the plane flown?

 4 A helicopter leaves its base and flies on a straight course for half and hour. The helicopter is now 57 km to the east of the base and 83 km to the north of the base.
  On what bearing did the helicopter fly?

 5 A hot air balloon flies 3.2 km on a bearing of 034°, and then 1.9 km on a bearing of 098°.
  How far east is the balloon from its starting position?

 6 Peter rows his canoe 1.3 km on a bearing of 310° and then rows 1.9 km on a bearing of 220°.
  On what bearing must Peter row his canoe to return directly to his starting position?

Unit 11

An introduction to probability

You will meet several words in this unit that you need to understand. Study the key vocabulary list carefully, making sure you know the meaning of all the words **before** you begin to study the maths. There are many other words, in addition to those in the list, that you might meet or find useful to use when studying this subject. Some are listed below.

doubt	even chance	fifty-fifty chance
good chance	likelihood	no chance
poor chance	predict	probable
risk	uncertain	unfair

The study of probability (also sometimes called chance) helps us to say in advance how likely it is that something will happen. So it is about guessing before something happens. The maths of probability does not give a certain answer unlike the maths we have studied before. **Statistics**, however, is the study of the facts and figures connected with actual events after they have taken place.

You will all have heard or seen a weather forecast. The people who prepare this report combine all the facts about the weather as it is now with all the recorded data about how the weather has behaved and changed in the past. This allows them to guess how today's weather will change and what it will be like tomorrow. They do not know for sure what will happen, but their guess is based on a lot of good information and so it is often correct. It is like making the best guess they can, based on the information they have.

Here are some of the types of weather that are possible:

sunny, cloudy, rainy, windy, stormy, snowy, . . .
(Can you think of some more?)

For any situation, all the possible things that could take place are called outcomes. If we are interested in one special outcome, it is called an event.

Think about each type of weather and decide what chance you think there is of this type of weather occurring in your town tomorrow. Here are some words you can use to help you describe these chances:

impossible   unlikely   equal (even)   likely   certain

For example, if you live in Bangkok, Thailand, the chance of snow tomorrow is unlikely. If you are in Greenland during January, the chance of snow tomorrow is **likely**. We might think that the chance of snow in Greenland is more than likely. We could say it is **very likely**. In the same way, the chance of snow in Bangkok is also **very unlikely**.

The probability scale

So far, we have used the words 'impossible', 'unlikely', 'equal', 'likely' and 'certain' to describe the chance that something will happen. We have also included 'very unlikely' and 'very likely'.
We can put these words in order to give us:

impossible   very unlikely   unlikely   equal   likely   very likely   certain

These words describe chances from 'definitely not, no way' all the way up to 'for sure'.

In maths we like to make things simpler by using numbers and symbols when we can.

- The number we choose to mean 'definitely not, no way' is 0.
- The number we choose to mean 'for sure' is 1.

This means that all the other words to describe a chance between 'impossible' and 'certain' would be represented by numbers between 0 and 1. We know that the numbers between 0 and 1 are fractions (or decimals).

This way of describing the chance or probability of something happening, using fractions or decimals between 0 and 1, is called the **probability scale**.

Fractions and decimals can also be written as percentages, so a probability can be written using fractions, decimals or percentages. For example, a probability that is described as 'equal' would be shown on the probability scale as $\frac{1}{2}$ or 0.5 or 50%.

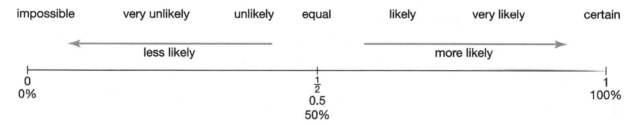

Many games are played by throwing (or rolling) dice to choose a number.
Dice are small cubes that have a different number of dots on each of the six faces, from 1 dot to 6 dots.

We will use dice in many of the examples in this unit. When we throw a dice, we cannot be sure which number will be facing upwards. There is an equal chance that it could be any one of the six numbers.

We know that a coin has two faces (or sides). The two sides of any coin look different, and we give them the names 'heads' (because a coin often shows a picture of a person's head) and 'tails'. The two sides of a coin are often used to choose between two different options, e.g. which team will kick off in a football match. One team chooses either 'heads' or 'tails' and the coin is then thrown up ('tossed'). The team that chose the side that is showing when the coin lands has 'won the toss'. If the other side of the coin is facing up, the other team wins the toss.

'heads'

'tails'

B Calculating simple probabilities

We use curly brackets, { }, to list all the possible outcomes, or the outcomes in a special event. For example:
If a normal dice is rolled, all possible outcomes would be:
Total = {1, 2, 3, 4, 5, 6}
If event *A* is 'the number thrown is even' then *A* = {2, 4, 6}.
If event *B* is 'the number thrown is bigger than 1' then
B = {2, 3, 4, 5, 6}.
If event *C* is 'the number thrown is prime' then *C* = {2, 3, 5}.

We can also write down the **number** of possible outcomes in an event. For the examples above we would write these as follows:

The total number of possible outcomes is written as n(*Total*) = 6.
The number of possible outcomes for event *A* is written as n(*A*) = 3.
The number of possible outcomes for event *B* is written as n(*B*) = 5.
The number of possible outcomes for event *C* is written as n(*C*) = 3.

Example

A bag contains one red sock, two green socks and four blue socks. I choose a sock at random from the bag.

i) List all possible outcomes.

We can use g1 and g2 to stand for the two green socks, and b1, b2, b3 and b4 to stand for the four blue socks.
Total = {r, g1, g2, b1, b2, b3, b4}

ii) List the outcomes for event *R*: 'the sock is red'.

R = {r}

iii) List the outcomes for event *G*: 'the sock is green'.

G = {g1, g2}

iv) List the outcomes for event B: 'the sock is blue'.

$B = \{b1, b2, b3, b4\}$

v) Write down the number of possible outcomes for each of the events R, G and B.

$n(R) = 1$, $n(G) = 2$, $n(B) = 4$

Simple probabilities are those where each outcome is equally likely to occur. For example, if we roll a fair dice, each of the six possible outcomes is equally likely.

In the example above, the socks are chosen at **random**, so each of the possible outcomes is equally likely. This is true for any event that occurs at random.

It is possible to calculate the probability for events like this, and we write the probability for an event A as P(A). Experiments show that for any event A:

$$P(A) = \frac{\text{Number of possible outcomes for the event } A}{\text{Total number of possible outcomes}} = \frac{n(A)}{n(Total)}$$

Example

Calculate the probability of each of the events in the previous example.

i) $P(R) = \dfrac{n(R)}{n(Total)} = \dfrac{1}{7}$

ii) $P(G) = \dfrac{n(G)}{n(Total)} = \dfrac{2}{7}$

iii) $P(B) = \dfrac{n(B)}{n(Total)} = \dfrac{4}{7}$

NOTE: Each probability is a fraction between 0 and 1, as we expect.

There are some **special probabilities** for the following events:

1 The event contains **all** the possible outcomes (i.e. it is certain). In our example about the socks, we could look at an event S: 'I choose a red, green or blue sock'. Then

$S = \{r, g1, g2, b1, b2, b3, b4\}$ and $n(S) = 7$

The probability of choosing a sock, $P(S) = \dfrac{n(S)}{n(Total)} = \dfrac{7}{7} = 1$

A probability of 1 on the probability scale stands for an event that is certain.

2 The event contains **no** possible outcomes (i.e. it is impossible). In the example of the socks again, we can look at the event Y: 'I choose a yellow sock'. Then

$Y = \{\}$ and $n(Y) = 0$

The probability of choosing a yellow sock, $P(Y) = \dfrac{n(Y)}{n(Total)} = \dfrac{0}{7} = 0$

> A probability of 0 on the probability scale stands for an event that is impossible.

3 Events that cannot happen at the same time.
Sometimes there are events that cannot happen at the same time. For example, if a coin is tossed, it cannot land on 'heads' and on 'tails' at the same time.
We call events like these mutually exclusive events.
If A and B are events that cannot happen at the same time, then we find that the probability of **A or B** is the same as the sum of the probability of A and the probability of B.

> If A and B are mutually exclusive events,
> $P(A \text{ or } B) = P(A) + P(B)$.

4 The outcomes for one event are all the possible outcomes that are **not** in another event.
If we start with all the integers from 1 to 20, $n(Total) = 20$.
Event A is 'choose a number that is prime'. So
$A = \{2, 3, 5, 7, 11, 13, 17, 19\}$ and $n(A) = 8$
$P(A) = \dfrac{n(A)}{n(Total)} = \dfrac{8}{20} = \dfrac{2}{5}$

All the possible outcomes **not** in event A will be all the other integers between 1 and 20. We call this set of outcomes **not** in event A, the complement of event A, written as A'.

So event A' is 'choose a number that is not prime'.
$A' = \{1, 4, 6, 8, 9, 10, 12, 14, 15, 16, 18, 20\}$ and $n(A') = 12$
$P(A') = \dfrac{n(A')}{n(Total)} = \dfrac{12}{20} = \dfrac{3}{5}$

$P(A) + P(A') = \frac{2}{5} + \frac{3}{5} = 1$

This is true for every event and its complement.

The sum of the probability of an event and the probability of its complement is equal to 1.

$$P(A') = 1 - P(A)$$

Example

We start with a normal pack of 52 playing cards. A card is chosen at random from the pack. Calculate the probability that this card is:

i) an ace

There are 4 aces in a pack.

So probability of choosing an ace $= \frac{4}{52} = \frac{1}{13}$

ii) a spade

There are 13 spades in a pack.

So probability of choosing a spade $= \frac{13}{52} = \frac{1}{4}$

iii) not a spade

Probability of choosing a card that is **not** a spade $= 1 - \frac{1}{4} = \frac{3}{4}$

iv) the ace of spades

There is only 1 ace of spades.

So probability of choosing ace of spades $= \frac{1}{52}$

v) an ace or a king

$P(ace) = \frac{1}{13}$. There are also 4 kings in a normal pack,

so $P(king) = \frac{1}{13}$.

As we choose only one card at a time, these two events are mutually exclusive.

So $P(ace\ or\ king) = P(ace) + P(king) = \frac{1}{13} + \frac{1}{13} = \frac{2}{13}$

 Exercise 1

1 Use one of the probability words: 'impossible', 'unlikely', 'equal', 'likely' or 'certain' to describe each of these events.

a) It is snowing somewhere in the world today.
b) It will rain for the next four days in your town.
c) A coin is tossed eight times and lands with tails up every time.
d) I roll a normal dice and get a score less than 7.
e) A coin is tossed and it lands with heads up.
f) I roll a normal dice and get an even number.
g) I choose a black bead from a bag that contains 2 white beads and 100 black beads.

2 Here are four events.

Event *A*   I roll a normal dice and get a four.
Event *B*   I roll a normal dice and get a nine.
Event *C*   I will die without oxygen to breathe.
Event *D*   There is a 60% chance I will pass my driving test.
Copy this probability scale and label each of the events marked on the scale.

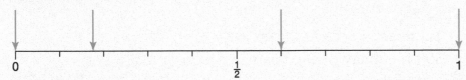

3 Here is a spinner.

a) List all the possible outcomes and write down n(*Total*).
b) Event *A* is 'the spinner lands on a multiple of 3'.
List all the outcomes of *A*, and write down n(*A*).
c) List all the outcomes of *A'*, the complement of *A*, and write down n(*A'*).

4 There are 36 chips in a box, numbered from 1 to 36. A chip is chosen at random.

a) Write down n(*X*) when *X* is the event 'a multiple of four is chosen'.
b) Write down n(*Y*) when *Y* is the event 'a prime number is chosen'.
c) Write down n(*Z*) when *Z* is the event 'a multiple of five is chosen'.

5 A fair dice is thrown. Work out the probability that the number is:

a) an odd number   b) bigger than 5   c) smaller than 3.

6 Here are some spinners. For each spinner work out i) P(white) and ii) P(grey).

a) 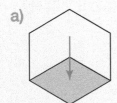   b) 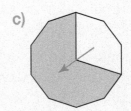   c) 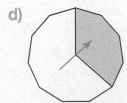   d)

7 The twelve numbered chips shown are in a bag.
Uncharee takes a chip at random from the bag.
What is the probability that Uncharee takes a chip:

a) with at least one 3 on it
b) that has not got a 3 on it
c) that has a 2 or a 3 on it
d) that has a perfect square on it?

33	34	35	36
37	38	39	40
41	42	43	44

8 A card is chosen randomly from a normal pack of 52 playing cards.
What is the probability that the card chosen is:

a) a black card
b) a club card
c) not a club card
d) the Jack of clubs

9 A letter is randomly chosen from the words 'INTERNATIONAL MATHEMATICS'.
Find the probability that the letter chosen is:

a) the letter T
b) a vowel
c) not a vowel.

Write these probabilities as percentages correct to 2 d.p.

10 A bag contains green, yellow and black marbles. A marble is chosen at random.
The probability of choosing a yellow marble is 0.26.
The probability of choosing a yellow marble or a green marble is 0.68.
What is the probability of choosing:

a) a green marble
b) a black marble
c) a red marble?

11 In a box there are 25 red dice and some blue dice. One dice is chosen at random.
The probability that it is a blue dice is $\frac{4}{9}$. Calculate the number of blue dice in the box.

12 Some black counters and some white counters are labelled either A or B as shown:

	A	B
Black	14	16
White	16	34

One of the counters is chosen at random.
What is the probability that this counter is:

a) a B
b) black
c) a black B?

A black counter is chosen at random.
d) What is the probability that this counter is a B?

A counter labelled B is chosen at random.
e) What is the probability that this counter is black?

C Combining events

So far, we have worked out the probability of only one event taking place at a time. It is possible for two or more events to take place at the same time. We can combine the results to work out the probabilities.

For example, if we toss two coins at the same time, the results for each toss will have two parts: one for each of the coins. All the possible outcomes will look something like this.

Coin 1	Coin 2
tails	tails
heads	heads
tails	heads
heads	tails

If we have many coins being tossed many times each, the number of possible outcomes is very large. We use two special sorts of diagram to show the outcomes clearly:

- a possibility diagram
- a tree diagram.

These will help us to work out the probabilities more easily.

Possibility diagrams

We can summarise the results of tossing **two coins** using a possibility diagram.

Let H stand for 'heads' and T stand for 'tails'. The possibility diagram would look like this.

		1st coin	
		H	T
2nd coin	H	HH	HT
	T	TH	TT

It is also possible to toss just **one coin twice** in a row. The possibility diagram for the results would look quite similar as there are also two possible outcomes for each throw.

		1st toss	
		H	T
2nd toss	H	HH	HT
	T	TH	TT

We will use these possibility diagrams to work out some probabilities.

Examples

a) Two identical fair coins are tossed.
  i) What is the probability of getting two tails?

  $Total = \{HH, HT, TH, TT\}, \quad n(Total) = 4$
  Probability of getting two tails $= \frac{1}{4}$
  ii) What is the probability of getting a head and tail?

  Probability of getting a head and tail $= \frac{2}{4} = \frac{1}{2}$

b) A fair coin is tossed twice in a row.
  i) What is the probability of getting two heads?

  $Total = \{HH, HT, TH, TT\}, \quad n(Total) = 4$
  Probability of getting two heads $= \frac{1}{4}$
  ii) What is the probability of getting a head and tail?

  Probability of getting a head and tail $= \frac{2}{4} = \frac{1}{2}$

c) A spinner has four equal sections that are numbered from 1 to 4. I spin twice and add the scores together.
  i) Find the probability of getting a score of at least 4.

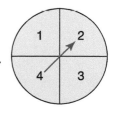

  The first thing to do is draw a possibility diagram to work out all the possible scores.

		1st spin			
		1	2	3	4
2nd spin	1	2	3	4	5
	2	3	4	5	6
	3	4	5	6	7
	4	5	6	7	8

From the possibility diagram we find that n(*Total*) = 16.
Probability of scoring at least 4 is $\frac{13}{16}$.

ii) Find the probability of getting a score of more than 4.

Probability of scoring more than 4 is $\frac{10}{16} = \frac{5}{8}$

iii) Find the probability of getting a score that is even.

Probability of an even score is $\frac{8}{16} = \frac{1}{2}$

d) Five counters numbered from 1 to 5 are in a bag. One counter is chosen at random and is not replaced in the bag.
Then a second counter is chosen from the bag at random. The two scores are added together.

i) Find the probability that both counters are even.

Draw a possibility diagram to work out all the possible outcomes.

		1st counter				
		1	**2**	**3**	**4**	**5**
2nd counter	**1**	✕	3	4	5	6
	2	3	✕	5	6	7
	3	4	5	✕	7	8
	4	5	6	7	✕	9
	5	6	7	8	9	✕

The first counter is not replaced before the second one is chosen. This means that it is not possible to have an outcome where both counters have the same number.
The diagram tells us that n(*Total*) = 20
The outcomes where both counters are even have been shaded pink. So the probability of this outcome is $\frac{2}{20} = \frac{1}{10}$

ii) Find the probability of scoring more than 5.

The probability of scoring more than 5 is $\frac{12}{20} = \frac{3}{5}$

iii) Find the probability of the score being even.

The probability of the score being even is $\frac{8}{20} = \frac{2}{5}$

Exercise 2

1 Here is a fair spinner with five equal sections numbered 2, 3, 4, 5 and 6.
Mary spins twice and adds her scores together.
Draw a possibility diagram to work out the probability of:

   a) scoring 3   b) scoring more than 7   c) scoring an even number.

2 Two normal fair dice are thrown together. Draw a possibility diagram to show all the possible outcomes for the two dice.
Find the probability that:

   a) the sum of the two numbers is 10 or more
   b) the two numbers are not the same
   c) at least one of the numbers is even.

3 Bag A contains counters numbered 2, 3 and 4. Bag B contains counters numbered 5, 6 and 7. A counter is chosen at random from bag A and then a counter is chosen at random from bag B. Draw a possibility diagram to show all the possible outcomes.
Find the probability that:

   a) the sum of the two counters is even
   b) the sum of the two counters is prime
   c) the product of the counters is even.

4 A fair dice is rolled and a fair coin is tossed at the same time. Draw a possibility diagram to show all the possible outcomes.
Find the probability of getting:

   a) an even number   b) a tail and a 3
   c) a head and an even number   d) a tail and a number less than 3.

5 Two spinners each have four equal sections. The sections on spinner A are numbered 2, 3, 5 and 7. The sections on spinner B are numbered 1, 4, 6 and 8. Each spinner is spun once.
Draw a possibility diagram to show all the possible outcomes.
Find the probability that:

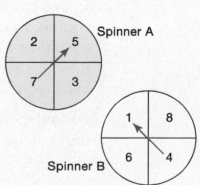

   a) one score is a factor of the other
   b) the sum of the scores is a prime number
   c) the sum of the scores is more than 10
   d) the sum of the scores is a perfect square.

6 Pupils in Year 9 must choose to do a project in two of these subjects: English, Science, Maths, Computer Studies.
Draw a possibility diagram to show all the possible pairs of subjects the pupils can choose for their projects. In some cases, one pair represents the same two subjects as another pair. Cross out one of these pairs.

a) Jay chooses both his projects randomly.
What is the probability that one of the subjects he chooses is English?
b) Sheena chooses Science and the other subject at random.
What is the probability that she chooses Science and Computer Studies?

7 Olga plays this game of chance at her town fair. The two spinners each have four equal sections. The sections on each spinner are marked with amounts of money to be won. Olga must pay 25¢ to play. She spins each spinner once and wins the total of the scores. Draw a possibility diagram to show all the possible outcomes. Find the probability that:

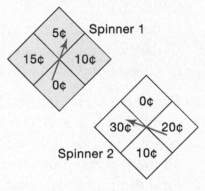

a) Olga wins more than 25¢
b) Olga wins 25¢ or less.
c) Would you play this game? Give a reason for your answer.

Independent events

If two dice are thrown, the outcome of the first dice has no connection with the outcome of the second dice. The two outcomes do not affect each other. We call events like these independent events.

I have counters numbered 1, 2, 3, 4 and 5 in a bag, and choose one counter randomly at a time. Suppose I don't replace the counters each time I choose one (similar to example d) on page 194). Now, the outcome of each choice depends on the outcome of the previous choices. These events are clearly connected and are **not** independent.

For independent events, the probability of **both** of them occurring is the same as the **product of their separate probabilities**.

$$\text{So } P(A \text{ and } B) = P(A) \times P(B)$$

This is true for any number of independent events.

$$P(A \text{ and } B \text{ and } C \text{ and } \ldots) = P(A) \times P(B) \times P(C) \ldots$$

Tree diagrams

A tree diagram can be used to show all the possible outcomes of events that are combined.

It can be used instead of a possibility diagram, and is especially useful to calculate probabilities when the outcomes of events are not equally likely.

The name of these diagrams indicates that they are made using different 'branches':

- Each separate outcome of an event is shown at the end of a different branch.
- The probability of each of the outcomes is written above or below the branch.
- The probability of an outcome following more than one independent event is found by multiplying the probabilities on each branch leading to that outcome.

Examples

a) A fair coin is thrown twice. Draw a tree diagram of the possible outcomes.

What is the probability of each outcome?

To work out the probability of each outcome, we multiply the probabilities along the branches for that outcome.

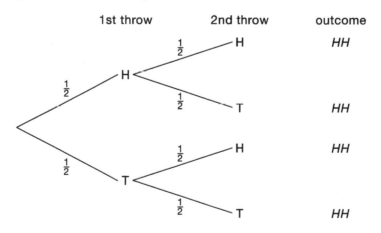

$$P(HH) = \tfrac{1}{2} \times \tfrac{1}{2} = \tfrac{1}{4}$$
$$P(HT) = \tfrac{1}{2} \times \tfrac{1}{2} = \tfrac{1}{4}$$
$$P(TH) = \tfrac{1}{2} \times \tfrac{1}{2} = \tfrac{1}{4}$$
$$P(TT) = \tfrac{1}{2} \times \tfrac{1}{2} = \tfrac{1}{4}$$

b) A fair coin and a normal fair dice are thrown at the same time.

  i) Draw a tree diagram to show all the possible outcomes.

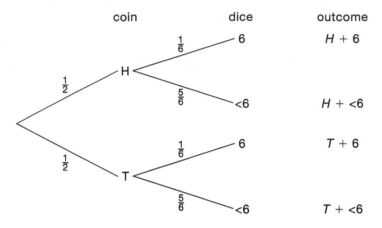

  ii) Find the probability of getting a tail and a six.

  Probability of outcome $T + 6 = \frac{1}{2} \times \frac{1}{6} = \frac{1}{12}$

  iii) Find the probability of getting a head and a number less than six.

  Probability of outcome $H + {<}6 = \frac{1}{2} \times \frac{5}{6} = \frac{5}{12}$

c) Bag A contains 4 black marbles, 3 white marbles and 2 yellow marbles.

  Bag B contains 3 black marbles and 2 white marbles.

  One marble is chosen at random from each bag.

  i) Draw a tree diagram to show the outcomes.

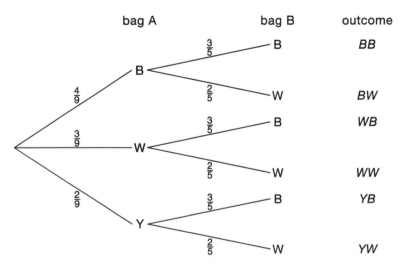

  ii) What is the probability that one marble is black and one is yellow?

  There is only one outcome with a black marble and a yellow marble.

  So probability of yellow/black $= \frac{2}{9} \times \frac{3}{5} = \frac{2}{15}$

iii) What is the probability that both marbles have the same colour?

The outcomes where the marbles are the same colour are:
BB or *WW*.
So probability of *BB* or *WW* = P(*BB*) + P(*WW*)

$$= \left(\frac{4}{9} \times \frac{3}{5}\right) + \left(\frac{3}{9} \times \frac{2}{5}\right)$$
$$= \frac{4}{15} + \frac{2}{15}$$
$$= \frac{2}{5}$$

d) In example d) on page 194 we considered a bag containing five counters numbered from 1 to 5.
One counter is chosen at random and not replaced in the bag.
Then a second counter is chosen from the bag at random.

i) Draw a tree diagram and use it to find the probability that both counters are even.

The bag contains five counters: two even counters (2, 4) and three odd counters (1, 3, 5).
Remember that after the first counter is chosen, the bag will contain only four counters, so the total number of possible outcomes for the second counter is only four.

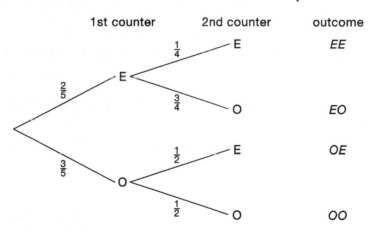

To work out the probability of each outcome, we multiply the probabilities along the branch for that outcome:

$$P(EE) = \frac{2}{5} \times \frac{1}{4} = \frac{1}{10}$$

The probability that both counters are even is $\frac{1}{10}$.

ii) Use the tree diagram to find the probability that both counters are odd.

$$P(OO) = \frac{3}{5} \times \frac{1}{2} = \frac{3}{10}$$

The probability that both counters are even is $\frac{3}{10}$.

iii) Use the tree diagram to find the probability that one counter is odd and the other is even.

$$P(EO) = \frac{2}{5} \times \frac{3}{4} = \frac{3}{10}$$

$$P(OE) = \frac{3}{5} \times \frac{1}{2} = \frac{3}{10}$$

The probability that one counter is odd and one is even

$$= P(EO) + P(OE) = \frac{3}{10} + \frac{3}{10} = \frac{6}{10} = \frac{3}{5}$$

Exercise 3

1 Paul buys six white chocolate bars and four milk chocolate bars. He chooses one bar at random and eats it. His friend Lee then chooses one bar at random and eats it.

  a) Draw a tree diagram to find all the possible outcomes.
  b) What is the probability that they both chose milk chocolate bars?
  c) What is the probability that they both chose different chocolate bars?

2 A man has some money in his wallet. He has four $10 notes, two $20 notes and two $50 notes. He takes out two notes at random from his wallet.

  a) Draw a tree diagram to find all the possible outcomes.
  b) What is the probability that the value of the two notes is:
    i) $20   **ii)** $30   **iii)** more than $50?

3 Abby writes a test in maths and in science. The probability that she passes maths is 0.6.
The probability that she passes science is 0.7. The two tests are independent of each other.
Calculate the probability that Abby:

  a) passes science and fails maths   **b)** fails both tests.

4 A spinner has sections of pink, yellow and blue as shown.

  a) Find the probability of it stopping on pink when it is spun once.

The spinner is now spun twice.
  b) Draw a tree diagram of all the outcomes.
  c) Find the probability that it stops first on blue and then on yellow.
  d) Find the probability that it stops on the same colour with both spins.

5 There are 4 girls and 12 boys that want to choose two teams. One child is chosen at random for Team A. Then one child is chosen at random for Team B from the children that remain.

a) Draw a tree diagram to show all the possible outcomes.
b) Find the probability that:
    i) Team A has a boy
    ii) Team A has a boy and Team B has a girl
    iii) both teams have a child of the same sex.

6 A group of 60 people consists of 24 females and 6 left-handed people. A person is chosen from the group at random.

a) Draw a tree diagram to show all the possible outcomes.
b) Find the probability that:
    i) the person is left-handed
    ii) the person is right-handed
    iii) the person is a female
    iv) the person is a male
    v) the person is a left-handed female
    vi) the person is a right-handed female or a left-handed male.

7 A teacher has a box that contains 16 new exercise books. 7 of them are green, 5 are yellow and 4 are black. One pupil picks a book at random. Then a second pupil picks a book at random.

a) Draw a tree diagram to show all the possible outcomes.
b) Find the probability that:
    i) both books are green
    ii) one book is green and the other is yellow.
A third book is chosen at random.
c) Find the probability that none of the three books is black.
    (Hint: Use P(*black*) and P(*not black*).)

8 Piet is a good football player. The probability that he will score in any match is $\frac{1}{5}$.
Piet plays the first two matches of the season.

a) Find the probability that:
    i) he does not score in either match
    ii) he scores in only one of the matches.
The probability that Piet scores more than one goal in any match is $\frac{1}{8}$.
b) Find the probability that he scores exactly one goal in the third match.
    (Hint: {score} = {score 1 goal + score more than 1 goal})

Unit 12 Statistical averages

Key vocabulary

arithmetic mean	data	measure of central tendency
average	frequency distribution	median
bimodal	mean	mode

In Unit 13 of Coursebook 2 we learned about collecting, organising and presenting facts and figures from everyday life, known as **statistical** data.

We looked at line graphs, pictograms, bar charts, column graphs, frequency tables, histograms, stem-and-leaf diagrams and pie charts. You may wish to revise that work before starting this unit.

Organising and presenting the data clearly gives us an overall picture of the information. To use this information effectively, we need to analyse it and make some decisions about what it means.

The first step in this process is to be able to work out some 'typical values'. This is rather like finding one value that will represent all of the separate data values we have collected. Sometimes the data values are all quite similar, but they can also be spread over a large range, so finding just one value to 'speak' for them all can be quite difficult.

Three different ways of choosing this single value have been defined. They are called statistical averages.

• The three statistical averages are the mean, the median and the mode.

> **1** If something is shared equally then the amount that each person gets is called the **mean**.
> **2** When the set of data is arranged in ascending or descending order, the middle value is called the **median**.
> **3** The value that occurs the most often is called the **mode**.

- These averages are usually somewhere in the middle (centre) of the given set of data and so they are also called measures of central tendency.
- These three average values are not often equal for a given set of data.
- Which of the three averages we decide to use at any time will depend on what the data is about and the sort of information we want to find out from the data.

 A ## Mean

The mean is the value that most people think of when we talk about an average. It is also called the arithmetic mean.

We find the mean by adding all the values in a set of data and then dividing by the number of values in the set.

$$\text{Mean} = \frac{\text{sum of values}}{\text{number of values}}$$

Examples

Find the mean of each set of data.

a) 34, 35, 36, 31, 29, 39

$$\text{Mean} = \frac{34 + 35 + 36 + 31 + 29 + 39}{6} = \frac{204}{6} = 34$$

b) 1, 1, 1, 1, 2, 2, 2, 3, 3, 4

$$\text{Mean} = \frac{1 + 1 + 1 + 1 + 2 + 2 + 2 + 3 + 3 + 4}{10} = \frac{20}{10} = 2$$

If the set of data is given in the form of a frequency table, then we may have to work out the total number of each item from the frequency data given.

For example, a supermarket buys 100 boxes of pears. It counts the number of rotten pears in each box and the results are shown in this frequency table. (It is also called a frequency distribution as it shows how the number of rotten pears is spread over the number of boxes.)

No. of rotten pears	0	1	2	3	4	5	6	7	8	9
No. of boxes	7	10	11	29	19	14	5	2	2	1

To calculate the mean number of rotten pears in a box, we must first find the total number of rotten pears in all the boxes. The data tells us that there are 7 boxes with no rotten pears. This gives 0 rotten pears so far.

There are 10 boxes that each have 1 rotten pear. This makes $10 \times 1 = 10$ rotten pears.

There are 11 boxes that each have 2 rotten pears. This makes $11 \times 2 = 22$ rotten pears (and a total of 32 so far).

We must continue like this with each number of rotten pears. These calculations are often added as another row (or column) to the frequency table/distribution.

No. of rotten pears	0	1	2	3	4	5	6	7	8	9	Totals
No. of boxes	7	10	11	29	19	14	5	2	2	1	100
Total no. of rotten pears	0×7 $= 0$	10×1 $= 10$	11×2 $= 22$	29×3 $= 87$	19×4 $= 76$	14×5 $= 70$	5×6 $= 30$	2×7 $= 14$	2×8 $= 16$	1×9 $= 9$	334

Now we can calculate the mean $= \dfrac{\text{total no. of rotten pears}}{\text{total no. of boxes}} = \dfrac{334}{100} = 3.34$

This figure does not mean that we have 3.34 rotten pears in every box. It is just a statistical value that tells us the average number of rotten pears in this set of boxes, and so it does not have to be a whole number.

Examples

a) The histogram shows the number of children per family in a group of families in a survey.

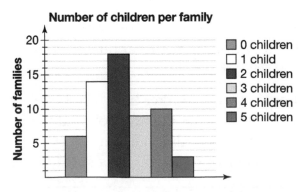

Calculate the mean number of children in a family.

First we must work out the total number of children.

No. of children	No. of families	Total no. of children
0	6	$6 \times 0 = 0$
1	14	$14 \times 1 = 14$
2	18	$18 \times 2 = 36$
3	9	$9 \times 3 = 27$
4	10	$10 \times 4 = 40$
5	3	$3 \times 5 = 15$
TOTALS	**60**	**132**

$$\text{Mean} = \frac{\text{total no. of children}}{\text{total no. of families}} = \frac{132}{60} = 2.2$$

So the mean number of children per family is 2.2.
Remember that this is a statistical value of the average; it does not mean that every family has 2.2 children.

b) The mean mass of a class of 30 pupils is 58.4 kg. A new pupil who weighs 63.6 kg joins the class. Calculate the new mean mass of this class.

$$\text{Mean mass of the pupils} = \frac{\text{total mass of pupils}}{\text{total no. of pupils}}$$

$$\begin{aligned}\text{Total mass of pupils} &= \text{mean} \times \text{total no. of pupils} \\ &= 58.4 \times 30 \\ &= 1752\end{aligned}$$

So the total mass of the class of 30 pupils was 1752 kg.

$$\text{New mean mass} = \frac{\text{new total mass of pupils}}{\text{new total no of pupils}}$$

$$= \frac{1752 + 63.6}{31} = \frac{1815.6}{31} = 58.6 \text{ (correct to 3 s.f.)}$$

So the new mean mass of the class is 58.6 kg.

c) A teacher keeps a record of the number of days each pupil is absent from school over a year.
This data is shown in the stem-and-leaf diagram.

No. of days absent

```
0 | 0 0 0 1 1 2 2 5
1 | 0 1 2 2 4 4 8
2 | 3 6 7
3 | 1 9                    Key:  2 | 3 = 23
```

i) How many pupils are in this class?
There are 20 results so there must be 20 pupils in the class.
ii) What is the least number of days a pupil is absent?
The least number of days absent = 0 days.
iii) What is the most number of days a pupil is absent?
The most number of days absent = 39 days.
iv) Calculate the mean number of days absent for this class.

$$\text{Mean} = \frac{\text{total no. of days}}{\text{total no. of pupils}}$$

$$= \frac{1+1+2+2+5+10+11+12+12+14+14+18+23+26+27+31+39}{20}$$

$$= \frac{248}{20} = 12.4$$

So the mean number of days a pupil was absent in this class is 12.4 days.

Exercise 1

1 Calculate the mean of each of these sets of data, correct to 1 d.p.

a) 1, 1, 2, 2, 3, 3, 3, 4, 5   b) 64, 81, 55, 71, 52, 73, 65, 67
c) 7, 5, 6, 9, 3, 7, 8   d) 6, 7, 8, 11, 13
e) 139, 165, 151, 149, 153, 145, 169, 136, 160   f) 3.2, 6.2, 3.6, 4.2, 5.4

2 The mean of six numbers is 53. Three of the numbers are 48, 45 and 54.
The other three numbers are all equal.
Find the value of these three numbers.

3 Six girls and four boys write an English test. The mean of their marks is 52. The mean of the six girls' marks is 53. Calculate the mean of the four boys' marks.

4 A garage employs seven qualified car mechanics and five apprentice mechanics. The mean monthly wage of the twelve mechanics is $800. The mean monthly wage of the apprentice mechanics is $660. Calculate the mean monthly wage of the qualified mechanics.

5 In a 200 m race, the mean time of the first four runners was 22.77 seconds. The winner finished in 21.47 seconds and this was 0.86 seconds faster than the runner who came second. Calculate the time taken by the third and fourth runners if they both finished at the same time.

6 A woman is practising for an archery competition. Each time an arrow hits the target she can score 1, 3, 5, 7 or 9 points. She shoots 60 arrows and her scores are shown in the table.

Score	No. of arrows
1	9
3	14
5	6
7	16
9	15

Calculate the woman's mean score correct to 2 s.f.

7 Some people who work in the centre of London were asked how long they spend travelling to work and back each day. The results of this survey are shown in the table.

Calculate the mean time taken to travel to work and back correct to 2 s.f.

No. of minutes	35	45	55	65	75	85	95
No. of people	4	6	10	15	8	5	2

8 Some teenagers were asked how many times they went to the cinema last month.
The results are shown in the histogram.

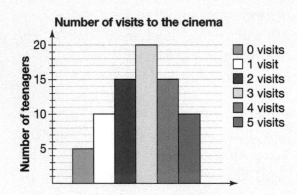

Number of visits to the cinema

Legend:
- 0 visits
- 1 visit
- 2 visits
- 3 visits
- 4 visits
- 5 visits

Calculate the mean number of visits to the cinema for these teenagers.

9 Kelly works on the express checkout at her local supermarket.
She kept a record of the number of items each customer bought during one hour last Tuesday.
Her results are shown in the histogram.

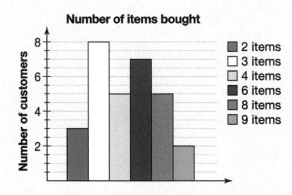

Number of items bought

Legend:
- 2 items
- 3 items
- 4 items
- 6 items
- 8 items
- 9 items

Calculate the mean number of items bought during this hour.

10 Abdul is testing a new feed for animals. He weighs some laboratory mice and records their masses. The mice then eat the new feed for a month and Abdul weighs them all again. The masses he recorded at the beginning and the end of his experiment are shown in the two stem-and-leaf diagrams.

Mass in g before		Mass in g after
7 7 9	1	8 9
0 1 2 2 4 6 8	2	1 1 2 3 5 8 9
0 0 1 2 3	3	1 1 2 2 3 4

Calculate the mean mass of the mice before and after the experiment to see if it is a good feed.

11 The school band is practising for a parade. The leader of the band keeps a record of the members who turn up for each practice session.
The results are shown in the stem-and-leaf diagram.

Number of pupils

```
2 | 4 6 7 7 8 8 9 9 9
3 | 0 0 0 1 1 1 2 2 2 3 3
```

Calculate the mean number of band members attending the practices.

B Median

Sometimes the mean value of a set of data does not really give us a 'typical' value.

For example, here are the heights of five friends: 145 cm, 153 cm, 151 cm, 172 cm and 149 cm.

$$\text{Mean height} = \frac{145 + 153 + 151 + 172 + 149}{5} = \frac{770}{5} = 154$$

The mean height of the five friends is 154 cm.

We can see that most of the friends (4 out of the 5) are actually shorter than this mean value.
The fact that one of the friends is especially tall makes the mean value not really 'typical' for this set of data.

If we arrange the heights in ascending order and then choose the middle value,

145   149   **151**   153   172

we find this gives the value of 151 cm, which is much more 'typical' for this set of data.

This value is called the **median**.

- The median for an **odd** number of values is the middle value when they are arranged in ascending or descending order.
- The median for an **even** number of values is the mean of the middle two values when they are arranged in ascending or descending order.

Examples | Find the median of each of these sets of numbers.

a) 2, 6, 5, 7, 3, 6, 4, 3, 5, 4, 2

Arrange the numbers in order:

2, 2, 3, 3, 4, 4, 5, 5, 6, 6, 7

There are eleven values, so we have one number in the middle:

2, 2, 3, 3, 4, **4**, 5, 5, 6, 6, 7

So the median of these numbers is 4.

b) 4, 2, 1, 3, 2, 4, 5, 6, 4, 1

Arrange the numbers in order:

6, 5, 4, 4, 4, 3, 2, 2, 1, 1

There are ten values, so we have two numbers in the middle:

6, 5, 4, 4, **4**, **3**, 2, 2, 1, 1

So the median will be the mean of these two middle numbers:

$$\text{median} = \frac{4 + 3}{2} = 3.5$$

Example | Here are the scores of some pupils in an exam, arranged in an unordered stem-and-leaf diagram.

Pupils' scores

```
9 | 6 3
8 | 5 9 4 3 5 7
7 | 3 5 7 9 8 3 4 2 1 1
6 | 8 6 3 2
```

Find the median score of these pupils.

When we learned about stem-and-leaf diagrams in Coursebook 2, we did not always put the 'leaves' in order.

If you are working out the median, you will need to put them in order.

Pupils' scores

```
9 | 3 6
8 | 3 4 5 5 7 9
7 | 1 1 2 3 3 4 5 7 8 9
6 | 2 3 6 8
```

There are 22 scores, so the middle two scores are 75 and 77.
The median score will be the mean of these two middle scores:

$$\text{Median} = \frac{75 + 77}{2} = 76$$

Example

The teams in a big football league all played a match last Saturday. The number of goals each team scored is summarised in the histogram.

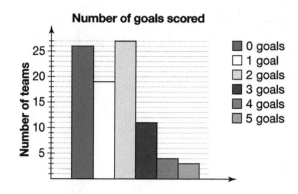

Number of goals scored

Legend:
- 0 goals
- 1 goal
- 2 goals
- 3 goals
- 4 goals
- 5 goals

Calculate the median number of goals scored by a team.

From the histogram, total number of teams
= 26 + 19 + 27 + 11 + 4 + 3 = 90
This means we have 90 sets of scores and the median will be the mean of the middle two scores in 45th and 46th places.
The first 26 teams scored 0 goals.
The next 19 teams all scored 1 goal. So the 45th team scored 1 goal.
The 46th team is in the next group and so they scored 2 goals.
Median score $= \frac{1 + 2}{2} = 1.5$

So the median number of goals is 1.5.

Exercise 2

1 Calculate the median of each of these sets of numbers.

a) 2, 3, 4, 5, 6, 4, 3, 2, 4
b) 77, 74, 79, 82, 88, 91, 71, 85
c) 5.1, 7.9, 3.6, 2.1, 7.9, 4.2, 3.6, 7.8, 3.6
d) $3.56, $2.79, $4.22, $6.83, $3.99, $2.84

2 The median of a set of ten numbers is $3\frac{1}{2}$. Nine of the numbers are 7, 1, 3, 12, 2, 5, 9, 1 and 3.
Find the tenth number.

3 The table shows the number of children in each family living on an estate.

No. of children	0	1	2	3	4	5
No. of families	6	5	7	4	2	1

Calculate the median number of children in a family on this estate.

4 The histogram shows the number of hours worked in a week by a number of factory workers.

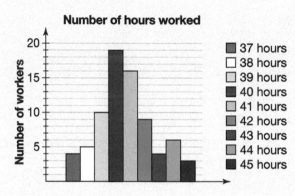

Number of hours worked

Legend:
- 37 hours
- 38 hours
- 39 hours
- 40 hours
- 41 hours
- 42 hours
- 43 hours
- 44 hours
- 45 hours

Calculate the median number of hours worked by these workers.

5 Jed and Fai are swimming lengths of their local pool. The times for each of them to complete the lengths are shown in the table.

Jed	30.0 s	29.7 s	30.1 s	29.6 s	42.3 s	30.0 s	29.7 s	30.1 s	29.9 s	30.1 s
Fai	30.7 s	30.1 s	31.8 s	31.1 s	30.1 s	31.7 s	30.6 s	30.1 s	31.1 s	31.5 s

Calculate the median time each of them takes and decide who is the faster swimmer.

6 The levels of calcium in some drinking water are measured in parts per million (PPM) for a number of towns.
The results are shown in the stem-and-leaf diagram.

Calcium level in PPM

```
1.0 | 4
0.9 | 8 1 1 9 0 3 2 2
0.8 | 4 2 8 7 7 3 4 8 4 3 3 5 1 7
0.7 | 6 5 1 9 8 7 9
```

Find the median of the calcium levels in these towns' water.
Give the answer correct to 3 s.f.

Mode

Retailers are interested in the item that is the most popular, the one they sell most of. Think about the size of pairs of jeans. Here are the sizes of the jeans that a shopkeeper sold in one day.

28, 30, 30, 32, 32, 34, 34, 34, 36

How can the shopkeeper decide what is the average size of his customers so he can stock more of this size?

He can work out the **mean** size of jeans bought by his customers:

$$\text{mean} = \frac{28 + 30 + 30 + 32 + 32 + 34 + 34 + 34 + 36}{9} = 32.2$$

This would indicate that he should stock most of size 32.

He can work out the **median** size of jeans bought by his customers:

Median = 32

This would also indicate that he should stock most of size 32.

Both of these averages do not tell the shopkeeper the size of jeans that **most** of his customers are buying. If we look at his sales figures, we find that more pairs of size 34 were bought in one day than any other size.

This is a much more useful statistical value for the shopkeeper. He knows that most of his customers buy size 34 and so he should make sure he always has more of this size in stock.

> The value that we find most frequently (or often) in a set of data is called the **mode**.

Examples

Find the mode of each of these sets of data.

a) 1, 5, 3, 8, 5, 4, 5, 9, 1, 5

The first thing to do is write the data in order (ascending or descending) so that it is easier to see which value appears the most often.

   1, 1, 3, 4, 5, 5, 5, 5, 8, 9

Now we can see easily that the most frequent value is 5.
The mode of this set of data is 5.

b) 15, 19, 14, 13, 20, 17, 16

   13, 14, 15, 16, 17, 19, 20

Each value in this set of data occurs only once, so there is **no** mode.

Example

Find the mode of this set of data: 52, 52, 64, 56, 58, 41, 49, 52, 45, 41, 69, 41.

Arrange the data in order: 41, 41, 41, 45, 49, 52, 52, 52, 56, 58, 64, 69
The values 41 and 52 both appear three times. They are the most frequently occurring values in the set.
So this set of data has two modes: 41 and 52.

NOTE: If a set of data has two modes, it is called bimodal.

D Choosing the mean, median or mode

a) The **mean** (properly called the **arithmetic mean**) is the most commonly used average to measure the central tendency of a set of data. This is the value that most people will think about when they are asked for an 'average' value for a group of numbers. It is a very useful and reliable measure of an average value in most cases.

However, if there are some 'extreme' values (values that are much bigger or smaller than the others in the set), then the mean value will be distorted and not give a true picture of the average. In these cases the median or the mode will usually represent the whole group better, because they are not affected by extreme values.

b) The **median** is the value usually preferred when discussing sociological and educational data. It is also commonly used in the analysis of general economic data. The main reason for this is that the data in these sections of science most often contain extreme values that would distort the calculation of the mean.

c) The **mode** as a measure of central tendency is most useful when dealing with the opinion or choices of people. As we saw earlier, it is the most sensible way for manufacturers or retailers to decide what to make in their factories or to stock in their shops.

These are general guidelines and are not meant to be 'rules' for choosing which average to use.

The final choice always depends on the details of the situation and what the average value is to be used for. So always think about the story involved and use your common sense.

Exercise 3

1 Find the mean, median and mode of each of these sets of data.

a) 3, 6, 8, 1, 8, 3, 3, 8, 6, 1

b) 10, 11, 13, 11, 15, 16

c) 110 g, 100 g, 120 g, 110 g, 100 g, 110 g

d) 2, 5, 6, 3, 7, 8, 4, 12, 11, 9, 10, 7, 6, 8, 9, 7

e) 12, 18, 24, 20, 18, 11, 20, 29, 41, 20

f) 9, 7, 4, 1, 3, 9, 9, 4, 0

g) 95, 96, 99, 98, 69, 59, 89

h) 5.1, 7.9, 3.6, 2.1, 7.9, 4.2, 3.6, 7.8, 3.6

i) 10.5, 9.6, 7, 11, 9.4, 8.1, 10.4, 11.7, 8.1, 9.4, 8.1

j) 2.8, 2.8, 9.3, 7.8, 6.9, 5.6, 2.0, 2.3, 6.9

2 For each of these frequency distributions, find i) the mean, ii) the median and iii) the mode.

a)

Value	2	4	6	8	10	12
Frequency	2	4	10	6	3	1

b)

Value	35	45	55	65	75	85	95
Frequency	4	6	10	15	8	5	2

3 Two normal fair dice are thrown together a number of times and the two scores added together. The results are shown in this table.
Find the mean, the median and the mode of the scores.

Score	2	3	4	5	6	7	8	9	10	11	12
No. of throws	1	1	3	4	6	8	3	2	1	1	0

Which statistical average do you think is the most sensible to use in this case?

4 A shopkeeper's records of how many cans of soda he sells each week are shown in the table.

No. of cans	32	57	82	107	132	157	182
No. of weeks	3	5	8	7	10	6	1

a) Calculate the mean number of cans of soda sold.

b) Find the difference between the mode and median values for this data.

5 The price of a popular brand of MP3 player was checked in a number of shops.

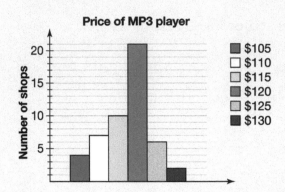

Price of MP3 player

Legend:
- ■ $105
- □ $110
- ■ $115
- ■ $120
- ■ $125
- ■ $130

The data from this investigation are shown in the histogram.
Find:

a) the mean price of the MP3 player
b) the median price of the MP3 player
c) the modal price of the MP3 player.

Which statistical average do you think is the most sensible to use in this case?

6 The ages of a group of people are 9, 11, 13, 13, 15, 15, 15, x, 18, 20. The median age is 0.4 bigger than the mean age.
Find a) the value of x and b) the modal age of this group.

7 The IQ levels of the pupils in one class are measured in a test.
The results are shown in the stem-and-leaf diagram.

```
12 | 0 9 4
11 | 1 5 4 1 3
10 | 6 5 7 3 7 4 6 7 1 2 5 0
 9 | 5 1 2 3 8 5 0 9 8 9 7 6 9
 8 | 6 5 4 8
 7 | 1 9 3
```

Find:

a) the mean IQ
b) the median IQ
c) the modal IQ for the pupils in this class.

Which statistical average do you think is the most sensible to use in this case?

8 The median of a set of eight numbers is $4\frac{1}{2}$. Seven of the numbers are 7, 2, 13, 4, 8, 2, and 1.
Find:

a) the eighth number
b) the mode of the full set of numbers.

9 Pupils in a Year 9 class in Flowerville School take part in an essay writing competition. The number of spelling errors that each pupil makes in their essay is recorded. The results are shown in the histogram.

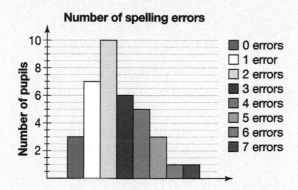

Find:

a) the mean number of spelling errors
b) the median number of spelling errors
c) the modal number of spelling errors.

Which statistical average do you think is the most sensible to use in this case?

10 The mean of four numbers is 54. The median of the same four numbers is 56, and their mode is 60.
Find:

a) the four numbers
b) the mean of the smallest and the biggest of the numbers.

11 For a biology project, some pupils measure the lengths of some leaves on a tree. Their results are shown in the stem-and-leaf diagram.

Length in mm

```
16 | 6 2 9 6 3 2 1
15 | 2 7 3 7 0 2 3 4 8 9
14 | 7 1 7 6 2 9 8 1 6 7 8 3 2 5 4 7 4
13 | 7 4 2 6 5 3 6 8 9 6
12 | 7 7 6 6
```

Find:

a) the mean length of the leaves
b) the median length of the leaves
c) the modal length of the leaves.

Which statistical average do you think is the most sensible to use in this case?